Frontiers in
HORTICULTURAL TECHNOLOGIES

THE EDITORS

Dr. Sanjai K Dwivedi, Scientist 'G' and Director DOP, DRDO, New Delhi is an alumnus of NDUAT Ayodhya (UP) and GBPUAT Pantnagar, (Uttarakhand). He has served at DIHAR, Leh (Ladakh), CEPTAM, Delhi, DIBER, Haldwani (Uttarakhand), and at DRL, Tezpur (Assam). He has made substantial contributions in remote border, cold arid and higher Himalayan regions, in the field of protected cultivation, and plant genetic resource management in indigenous and lesser known crops like sea buckthorn, cratagus, apricot, etc., for the benefit of armed forces, besides development of various products and technologies. His significant contribution includes the preparation, patenting and commercialization of herbal beverages of seabuckthorn. He also has experience in Human Capital Management in DRDO. Dr Dwivedi has over nine patents, over 91 research publications, 16 monographs and three books to his credit. He is a member of six professional societies, Editor-in-Chief of Progressive Horticulture Journal and Fellow of three professional societies. He is the recipient of the DRDO-Young Scientist Award, ICAR-Fakharuddin Ali Ahmed Award, ISHRD-Himadri Young Scientist Award and SAI National Senior Scientist Award.

Dr. Ranjan K Srivastava, is currently serving as a Professor in the Department of Horticulture, GBPUAT,Pantnagar. In his career spanning 25 years, he has worked on different aspects of Horticulture (Floriculture) including plant tissue culture and improvement, production technology and post-harvest technology. He has to his credit over 105 research papers, 7 books and 21 chapters. He has guided 22 M.Sc. and 12 Ph.D. students majoring in Horticulture (Floriculture & Landscaping). He has been a Member Editorial Board of Progressive Horticulture, Associate Editor and then Executive Editor of the International Journal of Basic and Applied Agricultural Research for over a decade, and Fellow of Indian Society for Horticulture Research and Development, Fellow, of the Confederation of Horticulture Associations of India, Fellow, Indian Society of Ornamental Horticulture. He has won Best Teaching Practices Award-2018of Pantnagar. Under Dr Srivastava's leadership, the Department bagged the Department Excellence Award 2017 and Best Horticulture Teacher Award of ISHRD. He was the organising Secretary of PHC 2023 Pantnagar.

Dr. S S Singh, is Presently working as Professor (Entomology) in GBPUAT, KVK, Almora. He is an alumnus of NDUAT Ayodhya (Uttar Pradesh). He started his career as Scientist (Entomology) in Department of Horticulture, Uttarakhand, before joining as Associate Professor at GBPUAT, KVK Dehradun (Uttarakhand). As Principal Investigator, he handled the project on integrated crop management in vegetable crops, litchi and apple funded by Horticulture Mission for North East and Himalayan States. Under his leadership, the KVK, Dehradun, has been awarded as Best KVK Award by Indian Council of Agricultural Research in the year 2013 and "Mahindra Samridhi India Agri Award in 2012 and 2017". GBPUAT, Pantnagar has also awarded KVK, Dehradun as Best KVK Award in Uttarakhand in 2010.

Dr Singh has over 62 research papers, 32 Scientific Articles, 6 Book Chapters, 12 Technical Bulletins and 3 Books to his credit. He was awarded as Outstanding Extension Scientist Award by GBPUAT in 2010 and 2015. He is honoured as a fellow of the Entomological Society of India, Society of Plant Protection Sciences, India Society of Horticultural Research & Development, Uttarakhand, Indian Society of Extension Education, New Delhi. He has organized four National Symposium on different aspects of horticultural sciences as organizing secretary in Uttarakhand. He is serving various professional societies in different capacities as General Secretary of India Society of Horticultural Research and Development, Uttarakhand.

Frontiers in
HORTICULTURAL TECHNOLOGIES

VOL. I

Sanjai K Dwivedi

Ranjan K Srivastava

S S Singh

2025

Daya Publishing House®

A Division of

Astral International Pvt. Ltd.

New Delhi – 110 002

ISBN: 978-93-5919-243-7 (HB)

Published by : **Daya Publishing House®**
A Division of
Astral International Pvt. Ltd.
– ISO 9001:2015 Certified Company –
4736/23, Ground Floor, Ansari Road,
Darya Ganj, New Delhi-110 002
Ph. 011-43549197, 8130496929
E-mail: info@astralint.com
Website: www.astralint.com

Preface

Globally, India is the second-largest producer of horticulture crops. The diverse agro-climatic and edaphic conditions prevailing in the country favours raising of a wide range of horticultural crops. The horticulture sector in India produces around 320 million tons of products and contributes about33% of the Gross Value Addition (GVA). The horticulture industry is becoming increasingly important to overall economic growth over time. As a diverse agricultural function, the horticulture industry has grown to be a promising source of employment possibilities. It is always desired that the current status must be assessed and based on the analysis of the present situation, a road map for the future course of action to meet out the prevailing and future challenges, needs to be chalked out. Considering the above-mentioned challenges of horticulture and their expected role in ensuring food and nutritional security to mankind by turning the Horticultural Sciences into useful technologies, a national conference 'Progressive Horticulture Conclave (PHC)'- 2023 was organized jointly by G. B. Pant University of Agriculture and Technology, Pantnagar and Indian Society of Horticultural Research and Development, Uttarakhand during February 3-5, 2023. The deliberations during PHC generated enormous interest in the horticulture fraternity and resulted in many useful recommendations for the consideration of different implementing agencies.

The Progressive Horticulture Conclave: 'Transforming Science into Technology ' was organized by the Indian Society for Horticulture Research and Development to act as a platform for researchers, policymakers, farmers, and students to exchange their ideas, innovations and research, as a part of the concerted efforts of the society towards achieving the goal. A galaxy of stalwarts, and policymakers of Indian Horticulture presented their insights on different aspects of horticulture. It was felt that a compiled document covering all these lectures would be beneficial for those associated with horticulture education, research, and development.

This book is the result of the selected discourse delivered at PHC 2023 Pantnagar by the eminent and lead/ invited speakers in the frontier areas of Horticulture. It is hoped that the compilation will be able to spread new ideas in the preparedness to face future challenges. The Indian Society of Horticultural Research and Development (ISHRD), Uttarakhand, and the editors are grateful to all the authors and the publisher for their valuable contributions and efforts made to realize this publication.

Sanjai K Dwivedi

Ranjan K Srivastava

S S Singh

Contents

Chapter 1

Entrepreneurship Development in Horticulture through Agribusiness Incubation

Maneesh Mishra* and Sharad Verma
Agri-Business Incubation Centre
ICAR - Central Institute for Subtropical Horticulture
Rehmankhera, P.O. Kakori, Lucknow – 226101 (India)
Email: abiu.cish@icar.gov.in / abi.cish@gmail.com

Beginning of 21st century is embarked with the rise of entrepreneurship. Paradigm of Indian agriculture has been shifted to a new dimension of Agri-preneurship. For nurturing innovations and start-ups in agriculture, various flagship programmes such as A Scheme for Promotion of Innovation, Rural Industries and Entrepreneurship (ASPIRE), Remunerative Approach For Agriculture and Allied sector Rejuvenation (RAFTAAR), Credit Guarantee Scheme for Startups (CGSS), NewGen Innovation and Entrepreneurship Development Centre (NewGen IEDC), Single Point Registration Scheme (SPRS), Modified Special Incentive Package Scheme (M-SIPS), Dairy Processing and Infrastructure Development Fund (DIDF), Multiplier Grants Scheme (MGS), Venture Capital Assistance Scheme (VCA), Prime Minister Employment Generation Programme and Other Credit Support Schemes, Marketing Promotion Schemes and Entrepreneurship Skill Development Programmes (ESDP) etc. are in force by the government of India (https://startuptalky.com/list-of-government-initiatives-for-startups/). Targeted cumulative efforts made India the 3rd largest startup ecosystem in the world (Economic Survey, 2022). Over 84,000 startups have been recognized by the Department of Promotion of Industry and Internal Trade (DPIIT) by the year 2022 (Press Release, 2022). Whereas in case of agritech startups, more than 450 were registered in 2019 in India, as per NASSCOM report (2019) which has been extended to 1,557 according to tracxn (https://tracxn.com/explore/AgriTech-Startups-in-India). The sector has raised close to $1.33 billion fund across 139 deals by 100 agritech startups between January 2020 and June 2022 (Upadhyay and Pathak, 2022).

Horticulture is contributing around 30 % in agricultural GDP of the country. India stands second in the world in fruit and vegetable production. Innovation led agri-businesses are coming up in pre to post harvest management, value chain and waste management. There exists a huge scope for innovative solutions in different domain of horticulture industry.

Current Status of Agri-Startups in India

Startups operating in agri-tech, horticulture, crops, food processing, organic agriculture, dairy farming, fisheries, farm mechanization, livestock and animal husbandry etc. are included under agri-startups. According to NAAS (2022), 1,485 agri-startups were recognized in 2020-21. Category wise ag-tech startups are given in **Table 1**. It is notable that only 48 startups belong to the horticulture category. If the food processing startups are also included under the horticulture, then the count goes to highest 1,822 among all.

Table 1: Number of Agri-startups in the year 2021

S. No.	Category of Agri-startups	Number of Agri-startups
1.	Fisheries	22
2.	Horticulture	48
3.	Animal husbandry and dairying	130
4.	Organic agriculture	474
5.	Food processing	1774
6.	Combination of such activities	74

Millet, a nutri-cereals, is now considered as the solution for global food and nutrition crisis. Identifying the nutritional value of millet, Indian Government has introduced the "India Millet Mission" to promote the millet production. Prime Minister of India inaugurated the Global Millets (Shree Anna) Conference on 18th March, 2023 in which he unveiled the Commemorative stamp and coin for millets (Press release, 2023). In the 75th session of the United Nations General Assembly in 2021, the year 2023 was declared as the International Year of Millets (IYM). More than 500 start-ups are working in the millet value chain and Indian Institute of Millets Research has incubated 250 start-ups under the Rashtriya Krishi Vikas Yojana – Remunerative Approaches for Agriculture and Allied Sectors Rejuvenation (RKVY-RAFTAAR) as per the Economic Survey (2023). With respect to the population involved in agriculture, number of startups in agriculture or horticulture sector is not proportionate. According to the States' Startup Ranking report (2019) only 4.2 % startups were recognized from the agriculture sector **(Figure 1).** Whilst, 11,417startups were registered over all for the same year.

In a report by NAAS (2022), Maharashtra stands to be the first in having agri-startup with highest 445 numbers followed by Karnataka and Uttar Pradesh with 285 and 194 startups, respectively. Delhi (175) and Gujarat (165) positioned fourth and fifth back-to-back with difference of only ten agri-startups. State wise DPIIT recognized agri-startups distribution for the year 2021 is presented in the **Figure 2.**

Results depict that first two states, Maharashtra and Karnataka accounted for half of the total agritech startups in the past five years. City wise number of startups registered represents that Bengaluru, Mumbai and Delhi NCR are the established startup ecosystem hubs and about 60% of the agri-startups come from the Tier I & II cities.

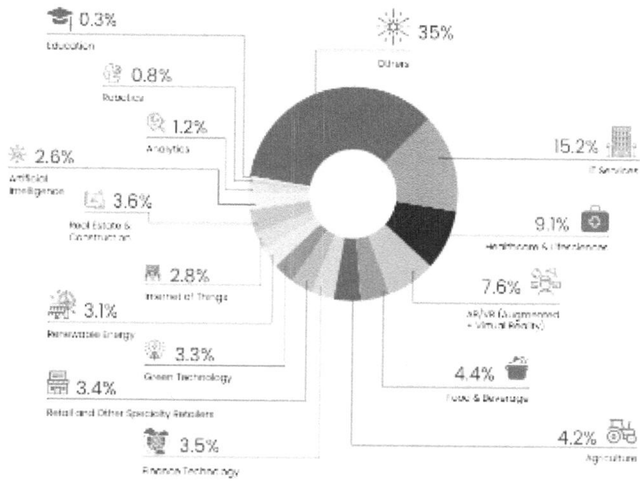

(Source: States' Startup Ranking, 2019)

Fig. 1: Sector wise categorization of recognized startups (2019)

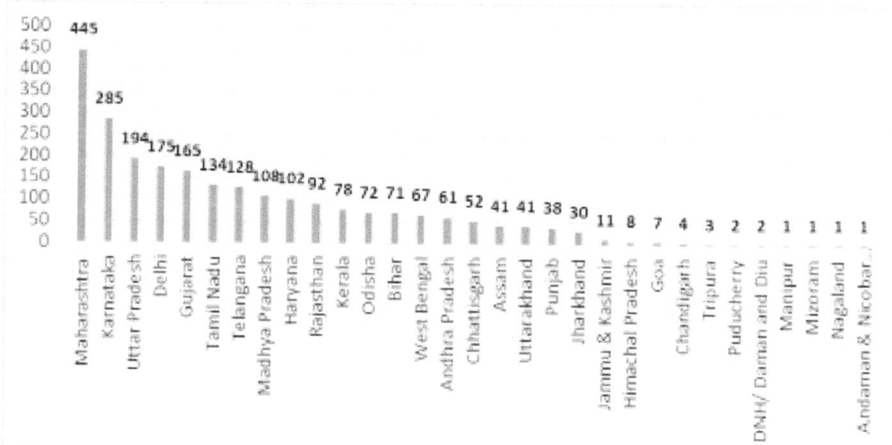

(Source: NAAS, 2022)

Fig. 2: State-wise number of DPIIT-recognized agri-startups in India (as of Nov, 2021)

Scope of Startups in Horticulture

Among agricultural sub sector, horticulture is of highest significance and contributes 30.4 per cent to GDP of agriculture alone (Economic Survey, 2014). India is the second largest producer of fruits and vegetables globally. Total horticultural production in the country was 311.71 million tonnes in 2017-18 (Horticultural Statistics at a Glance, 2018). However, there is enormous scope of expansion in horticultural sector. The productivity can be enhanced multiple folds and achievements may be extended to higher limits. Indian natural resources, land cover, diverse climatic zones, seasonal variations, biodiversity and manpower advocates far more growth returns from the horticulture system. Various components like production, practices, equipment, supply chain, post-harvest management and avoidance of wastage of stored produce needs to be upgraded with innovations. If these components are channelized through entrepreneurship, somehow, a significant growth may be achieved.

Role of Indian Council of Agricultural Research (ICAR) in Agri-preneurship

Despite, expeditious growth in entrepreneurship, there exist wide scope of growth for agtech startups in India. In this direction, the Indian Council of Agricultural Research (ICAR) is playing pivotal role in fostering the entrepreneurial ecosystem in agriculture unanimously. The organization is running 50 Agri-Business Incubation (ABI) Centres in its federated institutions. The ABI centres of ICAR are being operated under the World Bank-aided umbrella project "National Agricultural Innovation Fund (NAIF)"(Saxena *et al.*, 2020). The project is of prime concern among higher authorities of ICAR. They are very positive about the progress of the project and regularly monitoring to ensure the holistic support for its growth. A total of 818start-ups have been facilitated for agri-business related technical as well as infrastructural needs (Press Information Bureau, 2022) and 385 entrepreneurs/start-ups were graduated and trained for starting their business in agriculture and allied sectors by the council supported 50 ABI centers. ICAR-ABIs are meant for catering the business incubation support ranging from horticulture, spices, plantation crops, fisheries, rice, sugarcane, aquaculture, cotton, agricultural engineering, natural fiber, veterinary, dairy, meat, pig to post-harvest engineering etc. in respective institutes across the India. Thus incubation centres are catering key role in the development of agri-startups. Other than ICAR-ABI's several other incubation centres are run by the government and private organizations. Zone wise distribution of incubators is shown in the **Figure 3**.

Role of ICAR- Central Institute for Subtropical Horticulture in Horti-preneurship

The Central Institute for Subtropical Horticulture (CISH), Lucknow is one such ICAR institute operating Agri-Business Incubation (ABI) Centre in horticulture. The ABI centre of ICAR-CISH was inaugurated by the state agriculture minister Shri Surya Pratap Shahi Ji and horticulture minister Shri ShriRam Chauhan Ji by unveiling its brochure on 30th June 2020. The centre is thriving entrepreneurship in horticultural technologies ranging from production, protection and post-harvest management **Figure 4**. Institute has well developed laboratories with modern equipments and infrastructure. CISH horti-preneurs can use the institute's facilities for the development of products and prototype. They are provided with the mentorship by specialized scientists to develop desired technology or products.

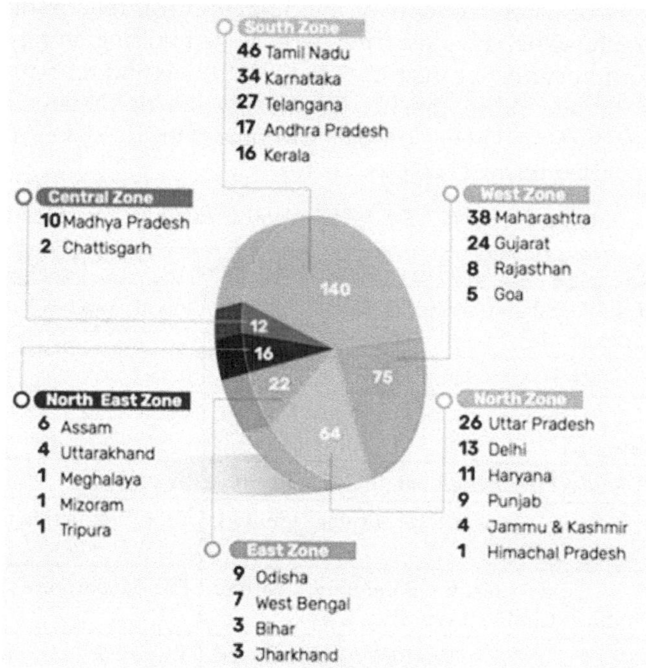

(Source: Bloomberg, Quint, 2021)

Fig. 3: Zone wise distribution of incubators

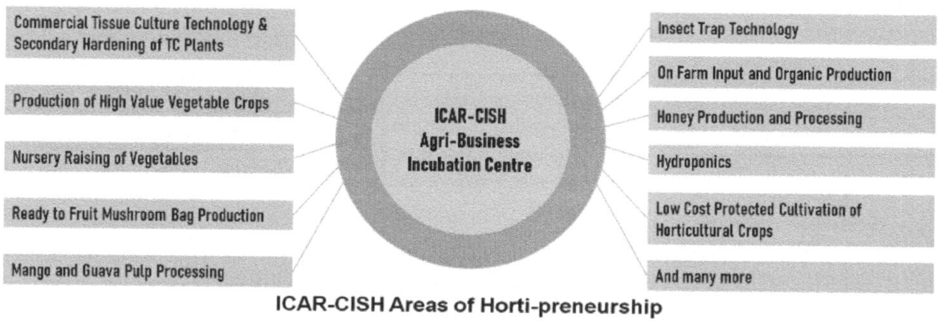

Fig. 4: Areas of Horti-preneurship in ICAR-CISH, Lucknow

The products can be tested at pilot and field level. After getting ready the products, the hortipreneurs are helped in test marketing, branding and promotion. They are also provided with the forward and backward linkages for their startups. ICAR-CISH supported hortipreneurs may use the facilities of office space/cubical, computer and internet in the campus. The CISH-ABI also offers need based services to any newly established startups or enterprises also. They may get assistance for product refinements, value addition, technical guidance and support. One can submit online application for incubation at https://cish.icar.gov.in/ or https://cish.icar.gov.in/abi_n.php

In effort of establishment of conducive entrepreneurship ecosystem in horticulture, ICAR-CISH is spurring early-stage budding entrepreneurs and has held up hortipreneurs in insect trap, value addition and food processing, digital marketing, tissue culture, mushroom, nursery raising, organic farming, supply chain management and honey traceability. Some of the ICAR-CISH incubated horti-startups are listed below **(Table 2)**:

Table 2: List of ICAR-CISH, Lucknow incubated horti-startups

S. No.	Name of startups / Firms	Technology
1.	M/s Parashar Agrotech Bio Pvt. Ltd., Varanasi, U.P.	Insect Trap
2.	M/s Life Speaks, Sangli, Maharashtra	Insect Trap
3.	M/s Ranaji Biotech India Private Limited, Kanpur, U.P.	Insect Trap
4.	M/s Saavi Industries, Sangli, Maharashtra	Insect Trap
5.	M/s Ranaji Biotech India Private Limited, Kanpur, U.P.	Crop Production
6.	M/s Apna Gaon Bioenergy Farmer Producer Co. Ltd., Ayodhya, U.P.	RTF Mushroom
7.	M/s Shristi Food Products, Malda, West Bengal	Value Addition and food processing
8.	M/s Jaishri Nursery, Varanasi, U.P.	High-tech Fruit Plants Nursery
9.	M/s Mango Baba, Lucknow, U.P.	Digital Supply Chain
10.	M/s Avakar EnviroSource Private Ltd., Lucknow, U.P.	Organic Farming
11.	M/s Balaji Agro Foods, Lucknow, U.P.	Value Addition and food processing
12.	*Farmers Family, Noida, U.P.	Supply Chain Management
13.	*Quantrello Digital Pvt Ltd., Kolkata, West Bengal	Honey traceability on BWS blockchain
14.	*Newtraway LLP, Lucknow, U.P.	Value Addition and food processing
15.	M/s Global Biotech & Agro Solutions, Lucknow, U.P.	Tissue Culture

ICAR-CISH, Agri-Business Incubation Centre is the best suitable place for emerging horti-preneurs to fulfill their necessities required to establish a startup. The incubation centre envisages promoting and nurturing young promising startups and horti-entrepreneurs with radical and innovative ideas in different streams of horticultural technologies from all over the country.

Agri-startup Dynamics

About 2,028 startups in agriculture are operational in India up-to-date and performing extremely well. It is noteworthy that highest numbers as well as most successful agri-startups are occupied in supply chain management which falls into the category of horticulture. Agri-startups have just gained the thrust in past two-three years. Statistics are changing with fast pace due to versatility in the agriculture sector. Ranking of agri-startups keep on rolling every year depending upon their fund raised, valuation, turn over or funding etc.

According to Entrackr report (2022), Ninjacart scored the first position in terms of valuation with 815 $Mn which was followed by the DeHaat (Farm services & products marketplace) and Absolute (Agri-biosciences) with equivalent 500 $Mn. Close to 500 $Mn, WayCool was also valued 460 $Mn startup. In the range of 200 $Mn, Arya (285 $Mn) and AgroStar (240 $Mn) were lined up. ReshaMandi, FarMart, Fraazo and Otipy were valued between 100 to 200 $Mn during the period of Jan 2020 to June 2022.

Revenue wise, Ninjacart, B2B supply chain startup reached the mark of Rs. 755 Cr leaving behind the DeHaat (Rs. 385 Cr), WayCool (Rs. 382 Cr) and Arya at Rs.196 Cr in the FY 21.

Steadily performing Ninjacart managed to get the highest funding of 145 $Mn for the period of 2020-22. Crossing the bench of 100 $Mn funding, Waycool, DeHaat and Absolute were posted with $117 Mn, $115 and $100 Mn, respectively. Besides, BigHaat, Unnati, Intello Labs, Bijak, Gramophone, Cropin, Krishify, VeGrow and BharatAgri also acted continuously in the space and raised themselves with a decent funding especially during the post Covid period.

Despite, speedy rise of agritech startups, none of them is able get entered into the unicorn club yet. However, Ninjacart was very close to touch the mark of unicorn in 2020-21 with the valuation of $815 Mn backed up by the Walmart, Flipkart and Tiger Global. Reasons for not yet having any unicorn in agriculture may be that agri-startups started getting large funds only in latest years. The sector has gained the attention of investors in recent years during Covid and post-Covid period when most of the business were closed except necessary services including agriculture and horticulture. Additionally, the agriculture and so its startups are ultimately dependent on the weather and variable season which is likely very uncertain factor. Regular climate change also brings ambiguity that keeps agri-startups on regular risk of loss. Low margins and high competition are also the reasons of slow returns in agriculture startups. Some of the agri-startups marking their consistent presence in the market are listed below in the Table 3:

Table 3: List of important Agri-startups in India

S. No.	Name of Agri-startups	Area of Agri-startups working
1.	Agrostar, DeHaat, Khetinext, Gramophone, LeanAgri, BharatAgri, BigHaat, A-One Seed, Terra Agro, AgriApp, SmartFarms, FarmGuru, Unnati, etc.	Upstream supply of quality inputs and farm solutions to the farmers
2.	CropIn, EM3Agri Services, Fasal, Hummin gbird, Trringo, FlyBird Innovation, Kritsnam Technologies, JaiKisan, Gramcover, Niruthi technology, Kheyti, Ecozen, SatSure, Farmguide, TartanSense, etc.	Precision farming through farm mechanization/ irrigation and digital technology

S. No.	Name of Agri-startups	Area of Agri-startups working
3.	PSamunnati, Jai Kisan, GramCover, Samaaru, PayAgri	Finance and insurance in the agriculture sector
4.	CropIn, Intello Labs, FarmERP, Jivabhumi, Agricx, qZense Lab, AgNext, Amnex, RML Agtech, Farm-Bee, Nebulaa's Matt, TartanSense, Yuktix, Fasal, etc.	Advisory, quality maintenance and traceability
5.	Animall, BigBasket, Ninjacart, WayCool, DeHaat, Bijak, Captain Fresh,JumboTail, Otipy, Kisan Net-work, Crofarm, Clover Ventures, Teabox, Tokri, Milkbasket, Farmpal, MeraKisan, VnF, FarmTaza, Green N Good, Organo fresh, Farmcon, Kamatan, Farm Fresh, etc.	Output market linkage (Downstream)
6.	Arya Collateral, Origo Commodities, Ergos, NCML, StarAgri, etc.	Warehousing and logistics
7.	Letcetra Agritech, Urban Kisaan, Future Farm, Bitmantis Innovation, Ela Sustainable Solutions, Acqua Farms, Junga FreshnGreen, etc.	Vertical farming/ hydroponics
8.	Licious, TenderCut, Fresh to Home, ZappFresh, Caprabook, Eggoz, Aquaconnect, Stellapps, Country Delight, Eruvaka, Aquaconnect, etc.	Livestock sector

(Source: Rao et al. (2020), ThinkAg (2021)

Areas of Entrepreneurship in Agriculture and Business Models

Opportunities are enormous for startups in agriculture because of the presence of the diversified areas of working for entrepreneurship. Chandra *et al.* (2020) have classified area of agri-preneurship / startups into eight following categories:

i. Output market linkages
ii. Facilitating input supply
iii. Farm mechanization and irrigation
iv. Financial solution (credit and insurance)
v. Quality maintenance through monitoring and traceability
vi. Post-harvest management
vii. Logistic services (warehousing & cold chains)
viii. Animal husbandry activities

These are the major gap in agriculture which offers the opportunities of entrance to the startups in the agri-industries. Various Agri-startups may work on these gaps of agriculture using different business models and provide solutions to users or farmers. Some common business models **(Table 4)** are given as follows:

Table 4: Various business models of agri-startups

Business Models of agri-startups	Example
Supply chain	E-distributor, Marketplace, Logistics, and Warehousing
Growing system and components	Aquaponics, Hydroponics, Vertical farming, Drip irrigation
Financial solution	Payments, Revenue sharing, Lending, Insurance
Farm data and analytics	Integrated Platform, Remote sensing software platforms, Farm mapping, Geo tagging, Farm management solution, Advisory
Field operations	Farm mechanization, Bulk handling

(Source: Chandra et al., 2020)

Modern Technologies in Agritech Startups

Agri-startups leverage innovations and technologies such as artificial intelligence (AI), Internet-of-Things (IoT), data analytics, machine learning (ML), software as a service (SaaS), data digitization, satellite data and blockchain for the development of their products or services. These technologies are used as tools by the ag-startups in their business models at various stages of value chain to solve the problems and improve the agriculture and its efficiency. Some of the technologies are detailed as below (https://inc42.com/resources/what-does-2023-hold-for-indian-agritech-industry/):

A. Artificial Intelligence (AI)

AI is the most utilized technology by the startups in agriculture. This enables the ability to replicate the human intelligence through artificial interface. Quick and automated results of AI attracted the agritech startups to leverage for addressing most of the agricultural problems. Easy handling of AI makes it to be effortlessly amalgamated in any system. Artificial intelligence has revolutionized the agritech industry with following range of applications:

i. **Supply chain management:** AI is widely exploited and successfully delivered its application in automated segregation and grading, washing, cutting, reduced spoilage, critical quality check, packaging, data transparency and maintenance, quality control and optimized logistics and transportation of high-tech supply chain management with enhanced efficiency and trusted transactions.

ii. **Quality control:** AI-powered cameras, drones and sensors have automated the real-time monitoring on various parameters of quality, data traceability, inspection of crops and reduction of waste etc.

iii. **Automation:** Labour-intensive tasks such as pesticide application, harvesting, fertigation, unreachable area of spraying and pruning etc. have been replaced by the automated AI-powered robots and drones which helped in improving the efficiency and reduction in costs incurred in agricultural practices.

iv. **Predictive maintenance and modeling:** Artificial intelligence has proven its worthy utility in predicting the reduced downtime and increased shelf life of post-harvest produce which allows its easy management and maintenance. AI enabled instruments are used in identification of disease outbreaks, determination of best time to harvest, crop yield prediction helping researchers and farmers to optimize operations and gaining increased yield of crops.

B. Blockchain Technology

Blockchain technology ensures the verification and authenticity of source of produce and tracking transactions facilitating the agri trade, food safety and quality standards. The technology has penetrated in global food chain ecosystem in recent years due to increased demand of traceability, data management and transparent supply chain.

C. Digitization of Food Quality

It is a new emerging technology revolutionizing the agri-tech industry. Technology involves the ability to scan the fruits and vegetables through computer vision and spectral sciences resulting into the quality and composition data. This may be proved very useful in post-harvest management and maintenance of agricultural produce.

D. Regenerative Agriculture

Regenerative agriculture is a holistic farming method that prioritizes the soil health by replenishing and enhancing resources instead of depleting them. This can be achieved by practices such as intercropping, providing diverse soil nutrients, crop rotation, cover cropping and limiting the use of synthetic fertilizers. This is highly application in improving soil health by retaining water and nutrients and reduced use of chemical inputs.

E. Internet of Things (IoT)

Application of IoT is used in precision agriculture, exclusively, by the agritech startups. IoT technology is utilized in sensors, drones and cameras for collecting the data on various biotic and abiotic parameters such as soil moisture, temperature, humidity, nutrient level, health and crop growth etc. which is then further utilized in big data analysis and machine learning.

F. Big Data Analysis

Data collected through IoT-enabled devices such as cameras, drones and sensors are further compiled and analyzed which help in optimization of operations and enhancing the accuracy and efficiency of actionable insights. Analyzed big data is then used in machine learning for creating programs and applications which facilitates prediction of real time attributes such as nutrient deficiencies, soil deficiencies, shelf life, fruit harvesting stage, ripening stage, identifying pests and diseases and degrading quality of food etc. Having real-time pre-delivered knowledge makes users able to take more informed decisions in businesses such as correct time of market release and pricing of the produce, accordingly.

G. Machine Learning (ML)

Agricultural data and knowledge collected through IoT is analyzed using machine learning algorithms to create any application such as detailed maps of the farm to optimize plant disease, characterization, irrigation, plantation and fertilization which may help in creating advisory, recommendations or device application etc.

H. Software as a Service (SaaS)

SaaS is a web-based software used to deliver applications over the internet as a service. This allows users to connect and use cloud-based device apps over the internet. The very common and most utilized examples in general are emails, cloud-based MS office which are operated at any place where the internet is available. In the same way, SaaS technology is used in agriculture also, making accessibility of agriculture applications. Though this technology is offering almost all operations in smart farming taking from digitization of food quality, farm management, finances, weather pattern prediction, crop cycle plan, crop diseases mapping, yield projection and capturing any real time farm data etc. However, the most popular known use of SaaS platform is creating transparency and traceability in supply chain. Microsoft Dynamics Agriculture ERP software is used in agriculture for crop cycle planning. One of the ICAR-CISH incubated horti-startups, Quantrello Digital Pvt. Ltd., Kolkata is working on honey value chain establishing transparency traceability and accountability using SaaS technology.

References

Anonymous, Chandra SRN, Srinivas K, Pandey N and Sharma R (2020). Startups with open innovation: Accelerating technological change and food value chain flows in India. Indian Journal of Agricultural Economics, Vol 75(4): 415- 437.

Anonymous (2014), Economic Survey. Agriculture and Food Management. In *Economic Survey of India, 2013-14* (pp. 137-159). New Delhi: Union Budget, Ministry of Finance, Department of Economic Affairs, Economic Division, North Block, N. Delhi.

Anonymous (2022), Economic Survey. Services. In Sanjeev Sanyal. Economic Survey of India, 2021-22 (pp. 213-342). New Delhi: Ministry of Finance, Department of Economic Affairs, Economic Division, North Block, N. Delhi.

Anonymous (2023) Economic Survey. Agriculture and Food Management: from food security to nutritional security. In *Economic Survey, 2022-23* (pp. 243-58). New Delhi: Ministry of Finance, Department of Economic Affairs, Economic Division, North Block, N. Delhi.

Anonymous (2022), Entrackr report. The rise of Indian agritech startups since 2020: Entrackr report. Agritech Report. (https://entrackr.com/2022/07/the-rise-of-indian-agritech-startups-since-2020-entrackr-report/).

Upadhyay, Harsh and Pathak, Shashank. (2022). The rise of Indian agritech starups since 2020 – Agritech Report. *Entrackr Report.*

Horticultural Statistics at a Glance (2018). Overview of Horticulture Sector. In Mamta Saxena. *Horticultural Statistics at a Glance 2018* (pp. 1-13). New Delhi: Horticulture Statistics Division, Department of Agriculture, Cooperation & Farmers' Welfare, Ministry of Agriculture & Farmers' Welfare, Government of India (www.agricoop.nic.in).

Anonymous (2022), NAAS. Agri-startups in India: Opportunities, Challenges and Way Forward, Policy Paper No. 108, National Academy of Agricultural Sciences, New Delhi: pp. 16.

Anonymous (2022), Press Information Bureau. Agritech Startups. *Press Information Bureau, Ministry of Agriculture & Farmers Welfare, Government of India.* Release ID: 1810911. Delhi. (https://pib.gov.in/PressReleseDetailm.aspx?PRID=1810911)

Press Release (2022). Sustained Government efforts result in increasing the number of recognized Startups from 452 in 2016 to 84,012 in 2022. Press Information Bureau, Ministry of Commerce & Industry, Delhi. Release ID: 1881495. (https://pib.gov.in/PressReleasePage.aspx?PRID=1881495)

Anonymous (2023), Press release. PM inaugurates the Global Millets (Shree Anna) Conference Unveils Commemorative Stamp and Commemorative Coin. *Press Information Bureau,* Prime Minister's Office. Release ID: 1908322. Delhi. (https://pib.gov.in/PressReleaseIframePage.aspx?PRID=1908322)

Saxena, Sanjeev, Datt, Shiv and Vikram Singh. (2020). Agri-business Incubation System in India: An Overview of ICAR. In Aruna T Kumar (Ed.). *Indian Farming* 70 (1), pp. 64-68. Pusa, New Delhi: Indian Council of Agricultural Research, Krishi Anusandhan Bhawan I, N. Delhi.

Maitra, Satyaki and Bhargava, Kabir (2019). More Than 450 Start-ups in the Indian Agritech Sector - Agritech Report. In Kavita Doshi. *National Association of Software and Services Companies (NASSCOM) – Agritech Report.* NASSCOM Press Office – Genesis BCW.

Anonymous (2019), States' Startup Ranking report. The Startup India Initiative. In *States' Startup Ranking 2019,* Department for Promotion of Industry and Internal Trade, Ministry of Commerce and Industry, GoI. pp. 29-48.

Anonymous (2021), ThinkAg. Ag-Tech in India: Investment Landscape Report 2021. ThinkAg, New Delhi.

Anonymous (2023), https://inc42.com/resources/what-does-2023-hold-for-indian-agritech-industry/

Anonymous (2023), https://tracxn.com/explore/AgriTech-Startups-in-India

Anonymous (2021), https://www.bqprime.com/opinion/the-geography-of-startup-incubation-in-india

Anonymous (2023), https://inc42.com/resources/what-does-2023-hold-for-indian-agritech-industry/

Chapter 2

Conservation of Horticultural Diversity at ICAR-NBPGR

Sandhya Gupta*, Vartika Srivastava and GP Singh
ICAR-National Bureau of Plant Genetic Resources, Pusa Campus, New Delhi-110012
**E-mail: sandhya.gupta@icar.gov.in*

India is endowed with huge genetic diversity of horticultural crops including vegetables, fruits, nuts, spices and medicinal and aromatic plants. The diverse agroclimatic conditions of India extending from temperate to arid, sub-tropical to tropical allows this diversity to flourish and perpetuate. Plant genetic resource (PGR) collections, which represent the genetic diversity of any crop or species, are essential for ensuring food security. With the advancement in science, new and easier technologies are constantly coming up to provide solutions for effective conservation of these important genetic resources. ICAR-National Bureau of Plant Genetic Resources (ICAR-NBPGR), is the nodal agency for the management of plant genetic resources in India alongwith its 10 regional stations (Fig 1).

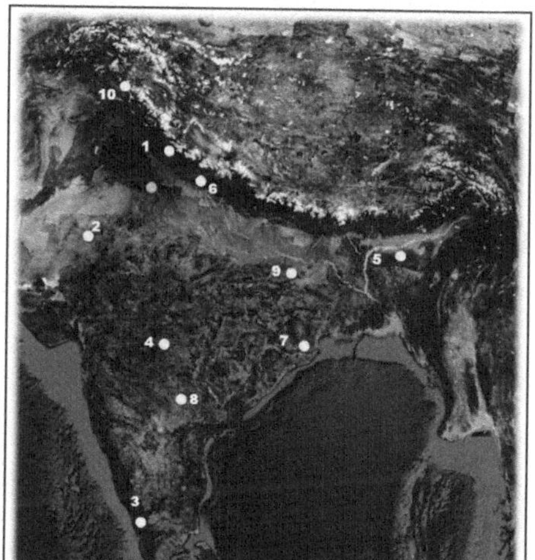

Fig. 1. Regional Stations of ICAR-NBPGR

Regional Stations

i. Shimla (Himachal Pradesh)
ii. Jodhpur (Rajasthan)
iii. Thrissur (Kerala)
iv. Akola (Maharashtra)
v. Shillong (Meghalaya)
vi. Bhowali (Uttarakhand)
vii. Cuttack (Orissa)
viii. Hyderabad (Andhra Pradesh)
ix. Ranchi (Jharkhand)
x. Srinagar (Jammu & Kashmir)

Strategy of Conservation of Horticultural Genetic Resources

Plant genetic resource conservation is critical as loss of biodiversity due to natural and anthropogenic factors is continuous and non-reversible. As per IUCN, a huge number of plants fall under endemic or threatened group, putting future sustainable development at risk. Many strategies are followed for effective management of PGR viz., *in situ* and *ex situ* conservation. In the *in situ* conservation, the diversity of plant species is conserved within its natural habitat in the form of biodiversity parks, sanctuaries and reserved forests. While in *ex situ* conservation, the plant part or propagules are conserved outside their natural growing conditions viz., as field gene bank, seed genebank, *in vitro* genebank and cryobanks. This involves significant costs, manpower, and energy consumption, making the strategy/process selection a top priority. Complementary conservation strategies are strongly advised to ensure effective and optimal conservation of a given species' maximum genepool. As a result, current germplasm conservation trends advocate preserving PGR diversity both *ex situ* (outside of their natural setting) and facilitating their continued evolution *in situ*, in their natural habitats.

The term 'gene bank' refers to an organizational unit tasked with managing PGR collections in order to conserve them and make their use more convenient. Gene banks are typically places where routine operations like managing plant collections in the field, storing seed, plant tissue, pollen, and/or DNA sequences are carried out. Thus, depending on the biology of the plant and the end user need, there are seed gene banks (SGB), *in vitro* gene banks (IVGB), cryo gene banks (CGB), DNA banks, or live collections in field gene banks (FGB) for conserving various types of PGR.

Ex situ Conservation at Field Gene Banks at NBPGR Regional Stations

Plant genetic resources are kept in field gene banks as live plants that grow and require maintenance on a continuous basis. FGB conservation is mainly used when the germplasm is either difficult or impossible to conserve as seeds (eg. seedless crops, recalcitrant seeded plant, or perennial species where seed production takes many years) or the crop is vegetatively reproduced.

Field gene banks allow for quick and easy access to plant genetic resources for characterization, evaluation, or utilisation. However, it is expensive to maintain because it require more labour, inputs, and space (land) than other methods of conservation. Also, plants maintained in FGBs are more vulnerable to natural disasters and adverse environmental conditions such as drought, floods, and pest and disease attacks.

Field gene banks are primarily used for the conservation of clonal crops, and are frequently used in conjunction with other conservation methods such as *in vitro* and cryo banks. Field genebanks are especially sensitive to germplasm health issues and regular monitoring and testing, as well as disease control measures, is required to keep plants disease-free. Although field gene banks are not the most secure method of germplasm conservation, they are often the only practical and cost-effective way to conserve clonal crop germplasm, particularly when resources and skills are limited. ICAR-NBPGR with its regional stations spread across the country is maintaining a good diversity of indigenous crops at FGBs as listed in the Table 1 (Annual Report, 2019).

Table 1: Important perennial horticultural crops being conserved in the field gene banks at the ICAR-NBPGR, Regional Stations

NBPGR Regional Station	Crops
Srinagar	Apple (*Malus* spp.), Pear (*Pyrus* spp.),Blackberry (*Rubus* spp.), *Allium* spp.
Bhowali	Kiwi (*Actinidia* spp.), Lemon (*Citrus limon*), Malta (*Citrus sinensis*), Citrus, Strawberry (*Fragaria xananassa*), Apricot (*Prunus armeniaca*), Pear (*Pyrus communis*), Plum (*Prunus domestica*), Peach (*Prunus persica*), Raspberry (*Rubus idaeus*), Walnut (*Juglans regia*), Apple (*Malus* spp.), Carambola (*Averrhoa carambola*), Grape (*Vitis vinifera*), Pomegranate (*Punica granatum*), Ficus (*Ficus benjamina*), Persimmon (*Diospyros kaki*), Cotoneaster (*Cotoneaster horizontalis*), Aonla (*Phyllanthus emblica*), Garlic (*Allium* spp.), *Oscimum*spp.
Shimla	Apple (*Malus* spp.), Peach (*Prunus persica*), Pear (*Pyrus* spp.), Plum (*Prunus domestica*), Apricot (*Prunus armeniaca*), Cherry (*Prunus cerasus*), Walnut (*Juglans regia*), Hazel Nut (*Corylus* spp.), Pecan nut (*Carya* sp.); Almond (*Prunus dulcis*), Kiwi (*Actinidia* spp.), Persimmon (*Diospyros kaki*), Pistachio (*Pistacia vera*),Ber(*Ziziphus mauritiana*), Hops (*Humulus* spp.), Citrus (*Citrus* spp.), Pine-apple guava(*Feijiosellowiana*), Fig(*Ficus carica*), Olive (*Olea europaea*), Hawthorn (*Crataegus* spp.), Chest nut (*Castanea* sp.),Viburnum Blackberry (*Rubus* spp.), Quince (*Cydonia oblonga*), Grape (*Vitis* spp.), *Mespilus* spp., Mulberry (*Morus* spp.), Pomegranate(*Punica granatum*), Rose (*Rosa* spp.) M&AP, Ornamentals
Shillong	Banana (*Musa* spp.), Citrus (*Citrus* spp.), Ginger (*Zingiber officinale*), Turmeric (*Curcuma* spp.), Taro (*Colocasia* spp.), Sweet potato (*Dioscorea* spp.)

NBPGR Regional Station	Crops
Jodhpur	Lemon (*Citrus limon*), Jamun (*Syzygiumcumini*), Pomegranate(*Punica granatum*), Ber (*Ziziphus mauritiana*) including "seed less landrace" of ber, Karonda (*Carissa carandas*), Carnauba(*Copernicia prunifera*), West Indian Cherry (*Malpighia emarginata*), Rohida (*Tecomella undulate*), Datepalm (*Phoenix dactylifera*),Lasura (*Cordia myxa*), Gundi (*Cordia rothii*), Drum stick (*Moringa oleifera*), Jangal jalebi (*Pithecellobium dulce*), Mulberry (*Morus alba),*Kair (*Capparis decidua*), Hingotia (*Balanites aegyptiaca*), Agave (*Agave americana*), Jojoba (*Simmondsia chinensis*), Guggul (*Commiphorawightii*), Anola (*Phyllanthus emblica*), Bael(*Aegle marmelos*), Guava(*Psidium guajava*), Custard Apple (*Annona squamosal*), Tamarind (*Tamarindus indica*), Aloe (*Aloe vera*), Phalsa (*Grewia asiatica)*
Ranchi	Jamun (*Syzygiumcumini*), Jackfruit (*Artocarpus heterophyllus*), Bael(*Aeglemarmelos*),Tamarind (*Tamarindusindica*),Aonla(*Phyllanthus emblica*), Lakoocha (*Artocarpus lacucha*), Banana (*Musa* species), Mango (*Mangifera indica*), Drumstick (*Moringa oleifera*)
Thrissur	Jackfruit (*Artocarpus* spp.), Mango (*Mangifera* spp.), Banana (*Musa* spp.), Malabar Tamarind(*Garcinia* spp.) and other minor fruits

Ex situ conservation at Seed Gene Bank

The best researched, most widely used, and most convenient method of *ex situ* conservation is storing genetic diversity as seed. Information pertaining to the storage requirements of major crop seeds is already known. Adequate moisture content before storage (seed drying to 3% for oily seeds and 5% or more for starchy seeds); appropriate storage temperature (-18°C is recommended for long-term storage); and careful seed production to ensure maximum longevity (storing them in hermetically-sealed containers). Seeds which tend to tolerate desiccation and freezing can be stored easily; however, recalcitrant seeds which are sensitive to desiccation are not amenable for seed genebank conservation.

Low temperatures and low moisture content are used to extend longevity and reduce regeneration intervals in orthodox seeds. Crop genebank requirements differ, and it is critical to choose a temperature and seed moisture content combination that is specific to the species and will maintain high viability for many years of conservation.

The National Gene Bank (NGB) stores germplasm as base collections at -18°C in accordance with gene Bank standards. It currently has 3,96,189 orthodox seed germplasm holdings representing 1,584 species including 151 species of vegetable (25,084 accessions), 35 (530 accessions) species in fruits, 661 species (6,771 accessions) of MAP and 17 species (3,721 accessions) of spices and condiments (Fig. 2).

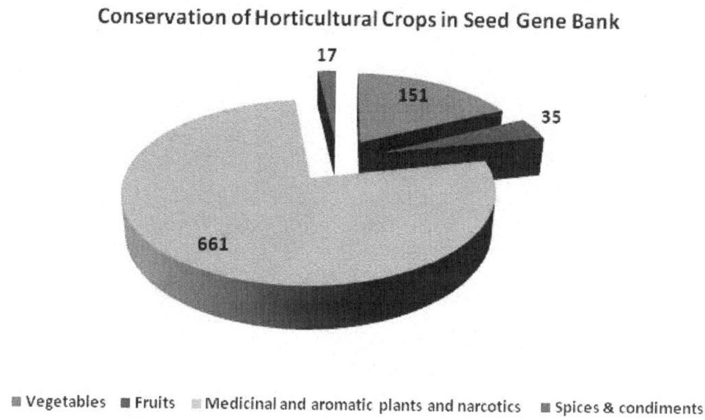

Fig. 2: Total number of species of different horticultural crops conserved as seeds in the Seed Gena Bank, ICAR-NBPGR

Ex situ conservation of Vegetatively Propagated Crops at *In Vitro* Gene Bank

In vitro conservation is the conservation of germplasm under artificial conditions in a relatively stable form on a defined nutrient medium in an *in vitro* gene Bank (IVGB). This technique is especially useful for clonally propagated species that are seed-sterile, produce highly heterozygous seeds, have short viability, or have long juvenile periods.

In vitro conservation, which involves maintaining explants in a sterile, pathogen-free environment, is thus preferred for clonal crop germplasm and the multiplication of species that produce recalcitrant seeds or do not produce seeds. It also promotes safe germplasm transfer under controlled phytosanitary conditions. Depending on the storage duration required, *in vitro* conservation methods differ. Medium-term storage (MTS) aims to slow growth and extend the duration of the subculture, whereas long-term storage (LTS) relies on cryopreservation, which is the storage of living tissues at ultra-low temperatures, typically liquid nitrogen (LN) (-196°C), for an indefinite period. Once stored in LN, germplasm is protected from contamination and requires minimal inputs. Thus, tissue culture techniques for medium-term storage (MTS) [In vitro active gene Bank (IVAG)] and cryopreservation techniques for long-term storage (LTS) [*In vitro* base gene bank (IVBG)] are used in the IVGB.

In vitro active gene bank

In the *in vitro* active genebank(IVAG), aseptic cultures are maintained under sub-optimal growth conditions with an objective to increase its storability in the controlled conditions thereby extending the subculture duration by slowing down or retarding the growth.

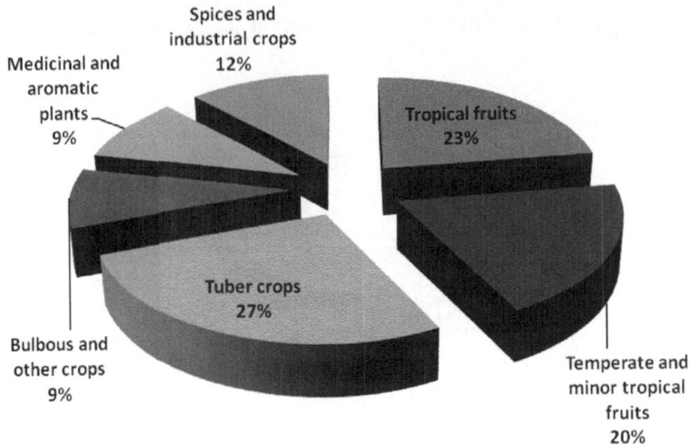

Fig. 3: Conservation status of germplasm collections of major crop groups in the *In Vitro* Gene Bank (November 30, 2022)

Slow growth strategies are based on extending the duration of subcultures without risking germplasm loss or jeopardising genetic stability through stressful treatments. The response of a species/genotype to growth-limiting treatments can vary greatly. There is no need to manipulate media or culture conditions for species that grow slowly naturally. To achieve slow growth in cultures under *in vitro* conservation, two or more techniques are usually combined. Viability, chlorosis, defoliation, browning, tip necrosis, hyperhydricity, and contamination must all be scored on stored cultures. Without subculture, cultures can last 6-30 months depending on the method used.The status of IVAG is depicted in Fig. 3.

In vitro base gene bank

The *in vitro* base gene bank ensures the preservation of vegetative propagules via cryopreservation, which is the non-lethal, viable, long-term storage of living tissues at ultralow temperatures (-196°C), of liquid nitrogen (LN). Plant cell metabolism is suspended at such low temperatures, eliminating the need to rejuvenate or regenerate the plant. Cryopreservation is the most widely used method of long-term storage of non-orthodox seeds, clonally propagated species, and cell lines.

Cryopreservation is simply the process of removing water from cells, which is important in preventing cellular damage caused by freezing injury. Cryopreservation can be attempted using either traditional (freeze-induced cooling in the presence of ice) or new cryopreservation (vitrification-based cooling in the absence of ice formation) techniques. Cryopreservation techniques for plant cells and tissues have been developed and tested over the last two decades and being followed for conservation of diverse horticultural species at ICAR-NBPGR as detailed below (Table 2).

Table 2: Status of *in vitro* base collection (as on March 31, 2022)

Crop/Species	Accessions	Technique(s)*	Explant (s)#
Allium spp.	155	V, DV	ST
Bacopa monnieri	6	V, DV	ST
Colocasia esculenta	2	DV	ST
Dioscorea spp.	15	V	ST
Fragaria spp.	2	ED	ST
Garcinia indica	1	DV	ST
Gentiana kurroo	3	DV	ST
Musa spp.	100	DV, V, AD	SM, ECS, ZE
Rubus hybrid	6	ED	ST
Vaccinium ovatum	7	ED	ST

**AD: Air dehydration; DV: droplet vitrification; V: Vitrification; ED: Encapsulation-dehydration; EV: Encapsulation-vitrifcation*

 # ST: shoot tip; SM: shoot meristem; ZE: Zygotic embryo; ECS: Embryogenic cell suspension

Ex situ conservation as Seed, Embryo, Embryonic Axes, Pollen and Dormant Buds at Cryobank

Horticultural crops of tropical region typically produce recalcitrant seeds, which are highly sensitive to desiccation and freezing. Such seeds need special conservation techniques for long-term storage without the loss of viability. For larger seeds, embryo or embryonic axes should be the preferred explants for cryo-conservation. However, certain species viz., mango, litchi etc produce highly recalcitrant seeds and in these crops, pollen cryopreservation has been standardized at ICAR-NBPGR. Pollen conservation is complementary to seed or vegetative propagule and hence pollen banks are generally established as supplements to the genebanks. Pollen storage is resorted mainly to serve plant breeders who attempt hybridization between plants cultivated/grown in different geographic regions, or showing non-synchronous flowering.

Vegetative buds in their dormant state are one of the best propagules to conserve and propagate identical clones of fruit crops and widely used as a routine procedure for cryopreservation of *Malus* spp., *Morus* spp., *Salix* spp. etc. The distinct advantages of dormant bud cryopreservation are maintenance of clonal fidelity, simpler as compared to vitrification and encapsulation methods of cryopreservation, long-term conservation is theoretically possible, dormant buds can be transported easily and regenerated via grafting or budding on the suitable root stocks. The status of germplasm conserved in cryobank at ICAR-NBPGR is as follows (Table 3):

Table 3: Status of germplasm at cryobank, ICAR-NBPGR(As on 30th Nov. 2022)

Categories	Accessions (no.)
Recalcitrant & Intermediate	
Fruits & Nuts	4188
Spices & Condiments	238
Plantation Crops	121
Agro forestry & Forestry	1645
Industrial crops	1353
Medicinal & Aromatic Plants (Orchid)	50
Total	7595
Orthodox	
Cereals	289
Millets and Forages	293
Pseudo-cereals	76
Grain Legumes	814
Oilseeds	682
Fibers	68
Vegetables	592
Medicinal & Aromatic Plants	1019
Narcotics & Dyes	35
Miscellaneous	79
Total	**3947**
Dormant buds	389
Pollen grains	642
Total	**12,573**
Genomic resources	2194
Total	**14,767**

(Source: ICAR-NBPGR, 2023)

Conclusion

The germplasm conservation of horticultural crops is difficult as compared to the field crops due to complexity in seed structure and physiology as most horticultural crops, particularly tropical fruit crops, bear recalcitrant seeds that arc difficult to conserve merely through desiccation and freezing. Many of these indigenous crops, such as mango, jamun, and garcinia, have huge diversity that must be conserved in FGBs in a systematic manner, as well as an appropriate alternative conservation strategy must be followed for their safety duplication. Research is needed to prioritize these species for developing long-term conservation protocols using new cryopreservation techniques. Furthermore, much emphasis is required to protect the crop's wild relatives for use in breeding programmes.

Compiled From

Annual Report (2019). ICAR-NBPGR. Online accessed in January 2023 (http.//nbpgr.ernet.in)

ICAR-NBPGR (2023). Website, ICAR-NBPGR (http.//nbpgr.ernet.in)

Chapter 3

Advances in Rootstock Research of Fruit Crops

O. P. Awasthi*, Theivanai, M. Kripa Shankar and Devachandra Nongthombam

Head, Division of Fruits & Horticultural Technology
ICAR–Indian Agricultural Research Institute, New Delhi-110012

Fruits are one of the major sources of micronutrients, phytonutrients, and dietary fibers that are essential for human health (Klee, 2010) and are often referred as 'Protective Food'. Owing to their commercial value and source of many processed food or secondary products, fruits play a key role in the economy of many developing countries. India is the second largest producer of fruits in the world, after China with a production of 107.10 million metric tons during 2021-22. However, productivity of fruit crops during the last one decade has been witnessing a declining trend from 17.6 Mt/Ha (2012-2013) to 15.3 Mt/Ha (2021-2022) thus resulting in a decrease of almost 13.06% which is thought provoking. The major factors attributed to productivity decline are several biotic (fungal, bacterial, viral diseases, insect-pests and nematodes) and abiotic stresses (salinity, alkalinity, drought, freeze/chilling injury). These stresses are the consequence of fruit crops being committed to the land for several decades with longer production periods and vulnerability to intermittent precipitation events, thermal stress, drought, salinity stress, indiscriminate use of irrigation water etc. Such stresses and further likely to aggravate under the era of climate change.

Rootstock selection and improvement with the goal of imparting tolerance to various biotic and abiotic stresses including canopy management for optimizing high density orcharding (HDO) would be an ideal strategy for mitigating these challenges. It is well understood that rootstock is inextricably linked with the success or failure of orcharding enterprise because plant's first line of defense against abiotic stress is in its roots. Certain species of rootstock and ecotypes seem to have a combination of key genes, quantitative trait loci and molecular networks that mediate plant responses to drought, salinity, heat, freeze tolerance and other abiotic stresses. Several studies have shown the importance of rootstocks for increasing food security by increasing the efficiency of natural (water soil) resource utilization and decreasing the use of chemical inputs, Albacete et al. (2015). Grafting is a millenary technique widely used in several different cultures, Mudge et al. (2009). Purposefully selected rootstocks enable the scion variety to express its genetic potential in terms of tolerance to various stresses, fruit quality and achieving real yield, modify architecture of plants. This book chapter is an attempt to provide a brief review of the rootstock research done in different fruit crops of national importance, technologies developed and future line of root stock research for the tropical, subtropical, semi-arid and arid fruits of the hot and cold region.

Mango

Mango (*Mangifera indica* L.) the national fruit of India and King of fruits is known as for its delicious taste, vitamins and minerals. It has a paramount place in Indian fruit industry accounting for 35% of area and 21% of total fruit production in the country and is also the largest producer of mango in the world which share 46% of the world mango production. However, mango production is simultaneously confronting various biotic and abiotic stresses due to emerging environmental conditions. These stresses often cause morpho-physiological, anatomical, biochemical changes, lead to the synthesis of reactive oxygen species (ROS), which can inactivate enzymes and damage cellular components ultimately affecting the growth and productivity of the crop.

To mitigate these challenges, the focus in mango rootstock research led to the development of rootstocks for varied situations that could confer resistance to adverse soil conditions and affect vigour of the scion. In this backdrop, aspects of polyembryony for uniformity; dwarfing characters for establishing HDP systems; tolerance to calcareous soils and physiological compatibility to scion assume specificity. For example, salt tolerant rootstocks viz., Bappakai, Olour, Kurakkan from India, 13/1 (Israel), Gomera 1 (Canary island), Sukary (Egypt) and GPL-1 and ML-2 tolerant to sodicity (Andaman Islands) has been identified (Gazit and Kadman, 1980., Duran Zuazo et al. 2003., Hafez et al. 2011). The salt-sensitive retort has been reported to vary among different mango rootstocks. Mango rootstocks such as "Olour" and "Turpentine" were reported to be saline tolerant, owing to their capacity to prevent the Cl—and Na+ ion uptake through enhanced proline concentration, Pandey et al. (2014). In another study, the rootstock of the genotype 'Gomera-3' exhibited more sensitivity towards salinity than the rootstock of the '13/1' genotype. The major physiological tolerance mechanisms exhibited by the '13/1' root stock include the increasing pattern of foliar K+ and proline concentration as well as the postponement of leaf abscission, Mahouachi et al. (2018). With the identification of these rootstocks, it is now possible to grow mango commercially in salt affected soils. Some tolerant rootstock such as Pullima for drought from Sri Lanka, flood tolerant rootstock, *Mangifera gedbe, Mangefera decandra,* Cultivar, Than ca (Vietnam), Turpentine and Number 11 for high soil pH (Florida) and rootstock tolerance to adverse drainage condition, Hilacha from Colombia have been found promising, thus offering scope for mango cultivation in these abiotic stressed areas (Jauhari et al. 1972., Van Hau et al. 2001., Mossler and Crane, 2013). Wild relative *Mangifera orophilla* has shown tolerance to cold.

Keeping in view the shrinking land resources, rootstocks such as Pusa Aam Moolvrant-1 and Pusa Aam Moolvrant-2 have been developed. PAM-1 rootstock genotype is able to reduce the plant height, canopy volume and diameter of mango variety Pusa Arunima, it reduces the tree vigour by more than 30% as compared to standard rootstock scion combination, Pusa Arunima-Kurukkan. PAM-2 is able to reduce the plant height, canopy volume and diameter of mango variety Pusa Arunima, Pusa Surya and Amrapali hence inducing dwarfism by reducing tree vigour by >20% in scion variety Pusa Arunima and >50% vigour in case of Amrapali and Pusa Surya as compared to standard combination of these scion with Kurukkan.

Citrus

Citrus is one of the important fruit crops of the world and is known as the "world fruit" for its high socio-economic and nutritional attributes. It is globally grown in areas with suitable climates between latitude 35 N~35 S. In India, citrus is the third most important fruit industry of India after banana and mango and share 15.53 per cent (1.09 million ha) of the total fruit acreage and contributes 13.8 per cent (14.81 million tonnes) of the total fruit production (2021-22). The main citrus groups in India are mandarins (*Citrus reticulata* Blanco), sweet oranges (*Citrus sinensis* Osbeck), lime (*Citrus aurantifolia* Swingle), lemons (*Citrus limon* (L) Burm. f), grape fruit (*Citrus paradisi* Macf.) and pomelo (*Citrus grandis* (L.) Osbeck). These citrus species are grown on array of rootstock for diverse morphological, physiological and biochemical characteristics. The significance of rootstocks in citriculture needs no emphasis because rootstocks have contributed perhaps more than any other factors to the success or failure of citrus crops. Rootstocks however, have their own merits and demerits for example sweet orange, grapefruit, mandarin and lemon on rough lemon rootstock are large, extremely vigorous and productive among most rootstocks' worldwide. However, scion cultivars on rough lemon are very susceptible to frost damage. Sour orange, although is an excellent rootstock for areas free of Citrus Tristeza Virus (CTV) but its susceptibility to CTV, particularly of sweet orange on sour orange has greatly restricted its use. *Jatti khatti* has been reported to be an ideal rootstock for Kinnow mandarin under arid and semi-arid conditions but it is somewhat susceptible to *Phytophthora*.

Due to long production periods and climate change, citrus production is impaired by wide range of biotic and abiotic stress. Among the various abiotic stress, phytophthora and huanglongbing (HLB) associated with *Candidatus liberibacter* asiaticus have resulted in a rapid decline of citrus orchards, while the abiotic stresses like salinity, drought and frost adversely affect the growth and productivity. Keeping in view the challenges confronting the citrus industry of India and across the globe, citriculturist all over the world however, have researched in depth and developed an array of rootstocks tolerant to various abiotic stresses. Citrus rootstock, Rangpur lime can adapt to alkalinity, salinity and calcareous soils and can impart resistance to cold and can provide high-quality fruits. Similarly, Cleoptera mandarin has the ability to tolerate salinity and cold (Santana-Vieira et al. 2016). F&A5 and F&A13 developed from Spain are tolerant to salinity (Forner et al. 2003). Karna Khatta (India) has been reported to be a drought tolerant rootstock for Kinnow mandarin (Kumar et al. 2018). Similarly, Sunki Maravilha mandarin (Brazil) can sustain drought. Sour orange the world most popular rootstock although has certain inherent problems is a cold tolerant rootstock and suitable for growing different citrus species in the colder regions. New generation/responsive rootstocks such as Alemow, X-639, Forner alcaide series of rootstocks, allotetraploid somatic hybrid rootstock Flohrag 1 were evaluated for tolerance to various degree of stresses. New breeding strategy of superior rootstocks "Super sour" and nucellar polyembryony have opened new vistas to breed superior varieties for tolerance to multiple traits.

Rootstocks that are seedy, highly nucellar, graft compatible with commercial cultivars, high degree of horticultural performance along with ensuring parity of fruit quality parameters with those of seedling origin have still remained a great challenge to the citrus industry. In this backdrop, citrus rootstock breeding objectives need to precisely focus on developing tolerance or resistance to biotic and abiotic stress factors viz., *Phytophthora* diseases, nematodes, salinity, etc., conform to horticultural performance requirements induced by the stock-scion interaction with respect to tree size, scion precocity, fruit size, dwarfness, yield and vigour. As prevalent elsewhere, rootstock scene in our country is also dominated by Sour orange, Rough lemon, Rangpur lime, Cleopatra mandarin and few others. Few interesting developments of the recent past are yet to be exploited commercially. Notable among them include, Citranges (*C. sinensis* x *P. trifoliata*); Citrumelos (*C. grandis* x *P. trifoliata*), Soost and Roose (1996); 'Carrizo', 'Troyer' citranges and 'Swingle' Citrumelo rootstocks are now widely used for harnessing the benefits of Phytophthora, virus and nematode tolerance.

Guava

Guava (*Psidium guajava* L.) is the fifth important fruit crop of India. It has gained considerable prominence on account of its high nutritive value, availability at moderate prices, a pleasant aroma and good flavour. It is one of the commonest fruits liked by the rich and the poor alike and is popularly known as 'apple of the tropics'. It is one of the hardiest fruit trees, adaptable to a variety of soil and climatic conditions and contributes 4% of the total fruit production in the country (2021-2022). The crop traditionally has been raised in many commercially producing areas through seeds even today resulting in variable tree and fruit characteristics. Dinesh and Vasugi, 2010 speculated that aneuploids, particularly trisomics, had promising qualities and may prove useful in developing plants with fewer seeds and dwarfing rootstocks. Sharma (1982) identified a promising tetrasomic dwarfing rootstock (Aneuploid No. 82), and studied its effect on growth and yield of 'Allahabad Safeda' which showed substantial dwarfing in terms of plant height, plant spread and tree volume. Overall yield/unit volume of the plant was highest in aneuploidy No. 82, which indicated its strong potential for use as dwarfing rootstock on a commercial scale, Sharma et al. (1992). In order to develop a rootstock tolerant/resistant to guava wilt and to test the possible role of different species, studies have been initiated on interspecific hybridization, in the genus Psidium. Leslie et al. (1995) and Edward & Shankar (1964) reported that *Psidium friedrichsthalianum* Niedenzu to be resistant to guava wilt. The other species reported to be resistant to the wilt are *P. cumuni*, *P. cattleianum* var. *lucidum*, *P. molle*, and *P. guineense* however, with contradictory reports by later researchers. Pathak and Ojha (1993) have also enumerated the potential use of these rootstocks and observed that Allahabad Safeda trees grafted on to *P. pumilum* had a dwarfing influence. *P. cujavallis* produced the largest trees but with non-uniform and rough skinned fruits. Since guava cultivation in many locations is threatened by wilt (*Fusarium* spp.), work on interspecific hybridization to develop wilt resistant rootstocks needs to be intensified. Efforts should also be made to develop wilt resistant scion varieties, of which, self-rooted plants could be used for commercial cultivation. The Philippine Guava (Purple type) has shown some promise as a wilt-resistant rootstock, but needs extensive experimentation. Extensive screening of other related *Psidium* species needs to be made for assessing their resistance to wilt. In this connection, reliable pathological screening techniques need to be standardized to hasten the process of disease resistance breeding in guava. The rootstock studies related to abiotic stress in guava are very meager. Sa et al. (2016) reported that in vitro screening of Brazilian guava rootstock 'Crioula' can tolerate salt concentration up to 1.8 dSm-1. Santhi et al., 2019 reported Mirzapur Seedling, Cheeni guava and Kerala as a potential rootstock for saline conditions. Other potential salt tolerant species as rootstocks include *P. cujavillis, P. mole, P. cattleianum* and *P. guineense*. In recent years due to increasing salinity levels and weather vagaries, nematodes have assumed a serious proportion. Guava rootstocks such as *P. friedrichsthalianum* (Chinese guava), *P. lucidum* (lemon guava), and *P. cattleianum* (strawberry guava) are resistant or immune to the root-knot nematode (M. enterolobii), Freitas et al. (2014). *Psidium guineense* showed tolerance to the nematode, Costa et al. (2012). Somatic cell genetic technologies for developing rootstocks tolerant to abiotic stress hold promise.

Grape

Grape is one of the important fruit crop grown on a variety of soil. This crop occupies sixth position amongst the fruit crop in India with a production of 1.21 million tonnes and contribute 3% of the total fruit production (2021-2022). Maharashtra is the leading grape producer in India (2022-2023). Till 1985 most of the grape vineyards were raised on their own roots as there were no noticeable problems due to the quality and availability of water used for irrigation. Subsequently, there were reports on reduction in yield due to soil and water related problems contributed mainly from the use of chemical fertilizers. The water used for irrigating grape was found with high reach of chlorides symptoms. Recurring droughts during the past few years have resulted in scarcity of irrigation water in Maharashtra and other parts of the country. Thermal stress, chilling/frost injury, mineral deficiency and toxicity are some of the other problems associated with the grape industry. The commercial harnessing of grape rootstock in context to production constrains has undergone sea change.

In India problems associated with grape production are salinity build up in soils and water, chlorides in irrigation water and excess levels of sodium and drought in and around grape cultivated area. Exploitation of rootstocks usage against salinity is linked to the ability to exclude moderate to high chloride and sodium exclusion combined with vigour imparting mechanism. The use of rootstocks in grape cultivation due to twin problems of drought and soil salinity has gained popularity, and almost all newer vineyards are being planted on stress tolerant rootstocks. Cultivated grape *Vitis vinifera* are highly susceptible to abiotic stress which can be efficiently controlled by using tolerant and resistant rootstocks of other Vitis species. The popular rootstocks 'Dogridge B' and '110 R' are presently being employed mainly to overcome the adverse effects of abiotic stresses like drought and soil salinity. *V. rupestris*, *V. berlandieri* and *V. champinii* wild types and its related genotypes are more tolerant to drought than *V. riparia* due to shallow root system. These species are important because it has the peculiarities of *Vitis rupestris* based rootstocks (St. George, 1103P, AXR#1 (3309C) which have broadly distributed roots, relatively drought tolerant, moderate to high vigor and mid season maturity. Similarly, *V. berlandieri* based rootstocks (110R, 140Ru, 420A, 5BB) are characterized by deeper roots, drought tolerant, higher vigor and delayed maturity. While, *V. champinii* related rootstocks are having (Dogridge, Ramsey (Salt Creek), Freedom, Harmony, GRNs) deeper roots, drought tolerant, salt tolerance but variable in hybrids. Pavlousek (2012) reported that rootstock hybrids that contain *Vitis berlandieri, Vitis riparia* and *V. cinerea* in their pedigree show better drought resistance as compared to hybrids with *Vitis rupestris* and *Vitis amurensis* which show a medium tolerance to these conditions. Jogaiah et al. (2014) reported that rootstocks belonging to *Vitis berlandieri* x *Vitis rupestris* (110R, 1103P, 99R, and B2/56) showed better osmotic adjustment and increased water use efficiency followed by the rootstocks belonging to *Vitis champinii* species (Dogridge and Salt Creek). Accumulation of high phenols in resistant rootstocks can be one of the indicators for screening rootstocks for abiotic stresses studied (Jogaiah et al., 2014). Plastic varieties such as cabernet Sauvignon contains higher level of aldehydes, IBMP and β-damascenone under arid climatic conditions (Xu et al., 2015). Rootstock selection should also focus on soil-borne pest presence (phylloxera, nematode). Grape phylloxera (Daktulospharia vitifoliae Fitch) is a gall forming aphid like that is the main soil pest of vineyards. Most rootstocks currently in use are only phylloxera tolerant rather than resistant. *Vitis riparia* besides being more cold hardy resists phylloxera. Another species *Vitis rupestris* of which St. George is an example also offers resistance to Phylloxera. Nematodes are the second important soil pest for the grapevines. It not only transmits viruses but also affect water stress, nutrient deficiency, stunting of shoots, and decreased yield. In this context, Dogridge has been found to be an excellent rootstock for controlling nematodes, simultaneously providing excellent nutritional support to the grafted scion. Since common grape rootstock have a narrow genetic base, efforts to extend the gene pools for breeding programs by using other species are of importance in the viticulture industry.

Underutilised Fruits

Covering an additional area on good lands for increasing production and productivity of fruit crops is rather difficult in the pace of ever increasing population pressure and declining land-man ratio. The alternative seems either to intensify the productivity on good lands by adopting modern tools and techniques or to occupy untapped resources for the purpose. Focusing to later approach, about 12.02 per cent land surface of the country is categorized as arid region, including cold arid region where there is ample scope for production of underutilized quality fruits because of the peculiar edapho climatic conditions. The important fruit crops grown in hot arid region are *Ziziphus* species, *Punica granatum, Cordia species, Capparis deciduas, Salvadora oleoides, Emblica officinalis, Grewia subinaequalis, Ficus carica, Aegle marmelos, Psidium guajava, Tamarindus indica* and multipurpose trees (MPT's) like *Prosopis cineraria* etc. The cold arid region comprises of fruit crops like the sea buck thorn, the wonder plant of the Himalaya and the store house of nutrients. The vegetation existing in hot arid region although are very hardy to various stresses, during the past two decade are facing the desire consequences of frost and chilling injury particularly in fruit crops like aonla and ber. The available rootstocks in these fruit crops have not been fully exploited for its tolerance to the various stresses. The important rootstock and its role in mitigating the abiotic stress in some of the arid fruits which also grows well on the marginal lands are discussed below.

Ber

Ber (*Ziziphus mauritiana* Lam) an apple of the desert have been substantially suffering from cold/frost injury causing economic loss to the ber growers. Maheshwari and Singh (1965) recogonised six economically important species in India. These are *Z. mauritiana* (ber), *Z. nummularia* (jharber), *Z. oenoplia* (makoh), *Z. rugosa* (suran ber), *Z. sativa* Gaertn (*Z. vulgaris* Lam. Kandiari) and *Z. xylopyrus* Hochst. Ex A. Rich. Tent. (kat ber). In *Ziziphus*, cytoplasmic accumulation of hexose sugars, scyllo-inositol and proline during drought stress may confer an increased ability to tolerate lower osmotic potential, stabilize DNA and membranes, and also ameliorate the deleterious effects of free radicals produced in response to the combination of high irradiance and water deficit. The exploitable attributes of some of the species are given in Table 1. Breeding for tolerance to sodicity, salinity although so far not attempted, Awasthi et al. (1995) reported that wild jujube *Z. rotundifolia* can be grown upto 60.5% ESP and 20.25 dsm^{-1} salinity which suggest the scope for future breeding work on this line. From the diverse germplasm maintained on *Ziziphus rotundifolia* at ICAR-CIAH, Bikaner, cultivars Tikadi, Dandan and Mahravali were identified for their tolerance to frost and low temperature (-3°C). The role of rootstock for low temperature tolerance in the scion cultivars and vice-versa needs to be investigated for their use in the breeding programme. Dasuanzao and Cuisuanzao from China have some characteristic abilities of drought resistance, leanness tolerance and adaptability and should be exploited in the arid and semi-arid regions of the Indian subcontinent.

Table 1 Exploitable attributes of some wild species of ber (Pareek, 2001).

Wild species	Exploitable attributes
Z. nummularia and *Z. lotus*	(i) Drought tolerance (ii) Extensive root system
Z. jujuba var. *spinosa, Z. mistol*	(i) Resistance to low temperature damage
Z. mauritiana var. *rotundifolia*	(i) Tolerant to frost and low temperature damage

Aonla

Indian gooseberry (*Emblica officinalis* Gaertn) is indigenous to India and therefore, its variability is widespread (Pareek and Sharma, 2009). The deep root system, reduced foliage, dormancy of fertilized fruitlets (April-June), coincidence of fruit growth and development with moisture availability make aonla an ideal plant for arid and semi-arid region. Aonla can be grown in marginal degraded soil. It can be grown in acidic to saline/sodic soil. Aonla germplasm collections have been made under AICRP on Arid zone Fruits by the coordinating centers. The credit for spread of aonla cultivation in the country goes to NDUAT, Ayodhya which has large assemblage of aonla genotypes. Seedling selections have been done from the natural variability prevailing in different regions of the country, mainly Uttar Pradesh (Pratapgarh) and Gujarat. Several promising cultivars have been developed. However, no systematic rootstock research has been initiated in this vitamin C rich fruit crop, except few initial screening of cultivars to graded levels of sodicity and salinity. The aonla seedling cultivar Chakaiya, Kanchan and Francis can be grown in sodic soil upto 35.0 ESP and in saline from 6-9 dsm^{-1} (Pathak and Pathak, 2001). Aonla which once made the desert dwellers roll in money have been extensively endangered by frost injury and is gradually becoming extinct. The selection Goma Aishwariya made form ICAR-CIAH regional station, Godhra is early and drought tolerant. A very close relative of *P. emblica* found in Deccan Plateau and certain drier forest pockets of southern India is *P. indofischeri*, presumably different from *P. fischeri* Pax., which already existed in Africa. The fruits of *P. indofischeri* have been observed to be less vulnerable to fungal spots than those of *P emblica*. Thus this species could be usefully exploited. The other related species is *P. acidus* (L.) Skeels syn. *P. distichus* Muell. Arg., *Cicca acida* Merr., *C. disticha* (L. (Otaheite gooseberry or Malay gooseberry or star gooseberry) having smaller trees with rather spreading crown.

Bael

The bael or Bengal quince (*Aegle marmelos* (L.) Correa) is an under-utilized fruit of Indian origin, valued for its nutritive and medicinal properties and is known as panacea of stomach ailments. Considerable variability exists in Uttar Pradesh, Bihar and West Bengal. From the existing variability some superior selections such as NB4, NB5, NB7, NB9, NB-16 and NB-17 have been made by NDUAT, Ayodhya. ICAR-CISH, Lucknow has developed two promising cultivars CISHB-1 and CISHB-2. Few promising cultivars such as Pant Aparna, Pant Sujata, Pant Shivani and Pant Urvashi has been developed by GBPUA&T, Pantnagar. Recently new cultivars such as Goma Yashi, Thar Divya, Thar Neelkanth and Thar Shristi have been developed by ICAR-CIAH, Godhra. The development of improved cultivars has spurred interest in the commercial cultivation of bael in different parts of country. Despite the hardy nature of bael tree in terms of tolerance to salinity, drought and other stresses, the information on rootstock is meagre. Wide variations however, have been reported for salinity tolerance in bael seedlings ranging from 3.5 dSm^{-1} (Toan 2012) to 10.5 dSm^{-1} (Shukla and Singh 1996b). The improved cultivars such as NB-5 have moderate levels of salt tolerance and can be grown up to 6 dSm^{-1} water salinity, Singh et al. (2018). These observations suggest inherent genetic differences for salt tolerance in bael and needs in depth investigation both in terms of salinity and drought. Since bael belongs to the family Rutaceae, citrus rootstock should be explored for tolerance to the various stresses. In this direction compatibility of bael on different citrus species was investigated by Bao, 2012. Research results showed that bael was able to tolerate salinity up to 2 g/l NaCl in nutrient solutions and good compatibility in grafting among bael rootstock and scion's *Citrus maxima, Citrus nobilis, Citrus sinensis* and *Citrus reticulata.*

Jamun

Jamun (*Syzygium cuminii*) is a popular fruit of India and is believed to be a boon for diabetic patients. It is presumed that India is the largest producer in global jamun production but the statistics is totally obscure. Although indigenous to India, systematic orcharding is still lacking. Diversity exist and distribution of some economic species of Syzygium are given in Table 2.

Table 2: *Syzygium* species of horticultural interest found in India.

Species	Distribution
S. cuminii	Indo-Gangetic plains of north India, Tamil Nadu in south, widely distributed
S. arnottianum	Western Ghats, the Nilgiris, Palni and Anamalai Hills
S. bracteatum	Western Ghats, Eastern/north-eastern India
S. operculatum	Grows wild in Nilgiri Hills of Tamil Nadu, Western Ghats
S. aqueum	Mainly in Assam, Sikkim and Meghalaya, Eastern/ north-eastern India
S. fruiticosum	Grows as an avenue tree, widely distributed

The work on germplasm collection in jamun is scanty. Being highly recalcitrant, it conservation protocol is yet to be standardised. Most of the germplasms is in field gene banks at institutes, like ICAR-CIAH, CHES Godhra, ICAR-CISH, Lucknow, ICAR-IIHR, Bangalore. Except few varietal developments in Jamun such as Goma Priyanka and Thar Kranti from ICAR-CIAH, Godhra, CISH, Jamwant from ICAR-CISH, Lucknow, Ra-Jamun by Bihar college of Agriculture, Bhagalpur and Konkan Bhadoli by Regional fruit research station, Vengrula, systematic rootstock research in this humble fruit is lacking. Although Jamun can grow well under salinity and waterlogged conditions there is dearth of literature for its tolerance to various stresses. Wong et al., 2018 reported that it can tolerate low levels of salinity and can ensure drought while Singh et al reported that it can be grown in alkaline soil up to a pH range of 10.5. Jamun crop is facing severe genetic erosion and extinction of its several species as a result of activities related to the urbanization and intensive agriculture. The genetic diversity of the related wild species is of particular value in search for sources of resistance to biotic and abiotic stresses.

Annonas

In Annona's various species can be used as rootstocks to which desirable species can be grafted. Seedling rootstocks of annona's are generally derived from heterogeneous open-pollinated plants. *A. reticulata* can withstand diverse ecological conditions and survive long dry periods. It is very useful as a vigorous rootstock. It is well adapted to unfavourable soil conditions and can grow in soils having pH 5.0 to 8.0. Because of its high tolerance to variable soil types, it is considered to be a good rootstock for cherimoya and sour sop. *A. diversifolia* also has similar ecological amplitude, but has been less widely used.

Sea Buckthorn

Sea buckthorn, a naturally growing thorny, deciduous bush, indigenous to cold arid region of Ladakh is an unexploited fruit crop. It is fast emerging as an important crop due to its nutritional and medicinal properties. Oleaster (*Elaeagnus* spp), belonging to the Elaeagnaceae family, is a medium sized hardy tree found growing in abundance in Ladakh and other cold arid regions of the country. The preliminary study conducted on the possibilities of using it as rootstock for sea buckthorn has shown encouraging results. Sea buckthorn scion budded on one-year-old rooted cutting of oleaster as root stock has shown good growth in initial three years. However, delayed incompatibility is yet to be studied. It seems that oleaster is a potential plant to be used as rootstock for sea buckthorn, since it has very deep and strong root system and fast growing habits. Moreover, it is easily propagated through hard wood cuttings and does not have sucker problem as faced in sea buckthorn. Use of root stock may help in shaping the canopy of sea buckthorn.

Initial studies showed that inter-generic budding in sea buckthorn is an effective method to reduce the thorn intensity and control profuse suckering. The budded scions have shown faster growth, larger leaf size and reduced number of thorns as compared to the control. The budded unions have not shown any sign of incompatibility or over/under growth even three years after budding. However, delayed incompatibility and fruit yield & quality need to be studied in future.

Future Outlook

Requirements for an ideal rootstock are many and no single rootstock possess all of the desired attributes. The future requirements of rootstocks should focus on:

- Survey, selection and evaluation of a large number of indigenous species in different fruit crops so as to explore their possibility to use as rootstocks under different agro climatic conditions.
- Understand the gene pool to identify taxa and a series of specific genotypes which can be of value as rootstocks.
- To develop rootstocks which can control the overall tree morphology and its vigour and to maximise its adaptation to different edaphic conditions.
- To standardize rootstocks in relation to productivity objectives, for use in orchard versus agro forestry systems and for use in adaptation to stress conditions.
- For basic research on rootstock genotypes for stress tolerance and in relation to the value of useful soil microflora such as mycorrhizae needs to be investigated.

- Development, introduction and testing of new clonal rootstocks for fruits where such rootstocks are not presently available.
- Development of complex hybrids through inter and intra-specific hybridization to develop more versatile rootstocks to increase their usefulness and adaptation with respect to compatibility, size control, precocity, productivity and resistance to biotic and biotic factors.
- Development of virus free material for all commercial rootstocks to reduce virus related incompatibility problems and to maintain sustainability in productivity and fruit quality. There is a need to refine the procedures for virus testing so that in future all the commercial rootstocks are available as virus free.

References

Albacete, A. 2015. Unravelling rootstock scion interactions to improve food security. *Journal of Experimental Botany*. 66 (8), 2211-2226

Singh, Anshuman., Sharma, P.C., Kumar, A. and Meena, M. D. 2018. Genotypic differences for salt tolerance in bael (*Aegle marmelos*) cultivars. *Indian Journal of Agricultural Sciences*. 88(3):435-441. DOI:10.56093/ijas.v88i3.78526

Awasthi, O.P., Pathak, R.K. and Pandey, S.D. 1995. Effect of sodicity and salinity on four scion cultivars budded on Indian jujube (*Ziziphus mauritiana*). *Indian Journal of Agricultural sciences*, 65, (5), 363-67.

Costa, S.R., Santos, C.A.F. and Castro, J.M.C. 2012. Assessing *Psidium guajava, P. guineense* hybrid tolerance to Meloidogyne enterologii. *Acta Hortic*. 959:59–66.

Dinesh, M. R. and Vasugi, C. 2010. Guava improvement in India and future needs. *J. Hortl. Sci.*, 5 (2): 94-108, 2010

Duran, V. H., Raya, A. M. and Aguilar, J. 2003. Salt tolerance of mango rootstocks (Magnifera indica L. cv. Osteen). *Spanish Journal of Agricultural Research*, *1*(1), 68-78.

Forner, J. B., Forner-Giner, M. A., and Alcaide, A. 2003. Forner-Alcaide 5 and Forner-Alcaide 13: two new citrus rootstocks released in Spain. *Hort Science*, *38*(4), 629-630.

Edward, J.C. and Shankar, G.1964. Rootstock trial for guava (*Psidium guajava* L.). *Allahabad Farmer*, 38:249-50

Freitas, V.M., Correa, V.R., Motta, F.C., Sousa, M.G., Gomes, A.C. M.M.and Carneiro, M.D.G. 2014. Resistant accessions of wild *Psidium* spp. to *Meloidogyne enterolobii* and histological characterization of resistance. *Plant Pathol. J. 63*, 738–746.

Gazit, S. and Kadman, A. 1980. 13-1 Mango rootstock selection. *Hort Science* 57, 81-87.

Hafez, O. M., Saleh, M. A., Ellil, A. and Kassab, O. M. 2011. Impact of ascorbic acid in salt tolerant of some mango rootstock seedlings. *Journal of Applied Sciences Research*. *7*(11), 1492-1500.

Jauhari, O. S., Teaotia, S. S. ad Upadhyay, S. K. 1972. Rootstocks studies in *Mangifera indica* L. *Acta Horticulturae.* 24, 107-109.

Jogaiah, S., Ramteke, S. D., Sharma, J. and Upadhyay, A. K. 2014. Moisture and salinity stress induced changes in biochemical constituents and water relations of different grape rootstock cultivars. *International Journal of Agronomy.*

Klee, H. J. 2010. Improving the flavor of fresh fruits: genomics, biochemistry, and biotechnology. *New Phytologist*, 187 (1): 44-56.

Kumar, S., Awasthi, O. P., Dubey, A. K., Pandey, R., Sharma, V. K., Mishra, A. K. and Sharma, R. M. 2018. Root morphology and the effect of rootstocks on leaf nutrient acquisition of Kinnow mandarin (*Citrus nobilis* Loureiro *Citrus reticulata* Blanco). *The Journal of Horticultural Science and Biotechnology.* 93(1): 100-106.

Leslie, R.W. Landrum, Dennis Clark, P. William and Jeff Brendecke. 1995. Hybridization between *Psidium guajava* and *P. guineense* (Myrtaceae), *Econ. Bot.* 49:153-161.

Maheshwari, P and U. Singh. 1965. Dictionary of economic Plants in India. ICAR, New Delhi.

Mahouachi, J. 2018. Long-Term Salt Stress Influence on Vegetative Growth and Foliar Nutrient Changes in Mango (*Mangifera indica* L.) Seedlings. *Sci. Hortic. 234*: 95–100.

Mossler, M.A. and Crane, J. 2013. Florida/Crop Pest Mangement Program. Profile: Mango. University of Florida. IFAS Extension. Publication #CIR 140.https://edis.ifas.ufl.edu/pi052.

Mudge, K., Janick, J., Scofield, S. and Goldshimdt, E. E. 2009. A history of grafting. Horticultural Reviews. 35, 437-93.

Pandey, P., Singh, A. K., Dubey, A. K. and Dahuja, A. 2014. Biochemical and Salt Ion Uptake Responses of Seven Mango (*Mangifera indica* L.) Rootstocks to NaCl Stress. *J. Hortic. Sci. Biotech.* 89: 367–372.

Pareek, O. P. 2001. *Fruits for the Future 2: Ber*. International Centre for Underutilized Crops, University of Southampton, Southampton, UK.

Pareek, O. P. and Suneel Sharma. 2009. *Underutilized Fruits and Nuts.* Aavishkar Publishers, Distributors, Jaipur, India. Vol. 1, pp. 366 & Vol 2, pp. 342

Pathak, R. K. and Pathak, S. K. 2001. Fruit production in problematic soil. *Indian Horticulture.* 58:16-22. Pathak, R.K. and Ojha, C.M. 1993. Genetic resources of guava. In: Advances in Horticulture (Vol I). Chadha, K.L and Pareek, O.P., (Eds.) Malhotra Public House,New Delhi

Pavloušek, P. 2012. Screening of rootstock hybrids with *Vitis cinerea* Arnold for phylloxera resistance. *Central European Journal of Biology.* 7(4): 708-719.

Sa, F. V. D. S., Nobre, R. G., Silva, L. D. A., Moreira, R. C., Paiva, E. P. D. and Oliveira, F. A. D. 2016. Tolerance of guava rootstocks under salt stress. *Revista Brasileira de Engenharia Agrícola e Ambiental.* 20 (12):1072-1077.

Santana-Vieira, D. D. S., Freschi, L., da Hora Almeida, L. A., De Moraes, D. H. S., Neves, D. M., Dos Santos, L. M., ... and da Silva Gesteira, A. 2016. Survival strategies of citrus rootstocks subjected to drought. *Scientific Reports*. 6: 38775.

Santhi, V.P., Ramesh Kumar , K., Gurusamy, J., Auxcilia and Parthiban, S. 2020. Collection and Evaluation of Guava Germplasm for Physiological and Biochemical Basis for Sodicity Tolerance. Paper presented on National Conference Climate smart agriculture for livelihood security: challenges and opportunities 13-14, September 2019 held at Anbil Dharmalingam Agricultural College and Research Institute. Souvenier. pp :55

Sharma, Y.K. 1982. Rootstock investigation in guava (*Psidium guajava* L.). Thesis submitted for the award of Ph.D. degree to Meerut University, Meerut.

Sharma, Y.K., Goswami, A.M. and Sharma, R.R. 1992 Effect of dwarfing aneuploid guava rootstock in high density orcharding. *Ind. J. Hort*. 49: 31-36

Shukla, S. K. and Singh G, N. 1996b. Seed germination and growth of bael (*Aegle marmelos* Correa) seedlings as influenced by salinity. *Annals of Arid Zone.* 35: 385–6.

Soost, R. K. and Roose, M. 1996. "Citrus". In *Fruit breeding: Tree and tropical fruits,* Edited by: Janick, J. and Moore, J.N. 257–323. New York: Wiley.

Toan Nguyen Bao. 2012. Evaluation of salt tolerance (NaCl) and compatibility of bael [*Aegle marmelos* (L.) Corr.] as rootstock of citrus trees *TAP CHI SINH HOC*, 32(1). DOI:10.15625/0866-7160/v32n1.654

Van Hau, T., Viet Khoi. N. and Ngoc Tran, V. (2001). Effect of rootstocks on the growth of 'Cat HoaLoc' mango scion. Mini-symposium on the Activities of Subproject B2, CTU, 24-25.

Wong, A. 2018. The Art of Botany: The Effect of Drought on Plant Anatomy. *Horizons*. 3: 19.

Xu, X. Q., Liu, B., Zhu, B. Q., Lan, Y. B., Gao, Y., Wang, D. 2015. Differences in volatile profiles of cabernet sauvignon grapes grown in two distinct regions of China and their responses to weather conditions. *Plant Physiol. Bioch.* 89: 123–133. doi: 10.1016/j.plaphy.2015.02.020

Chapter 4

Challenges and Opportunities in Horticultural Nursery Technologies

Anil Kumar Dubey, Nimisha Sharma, Lal Chand and Mukesh Shivran
Division of Fruits and Horticultural Technology,
ICAR-Indian Agricultural Research Institute, New Delhi-110012 (India)

The diverse agro-ecological region of India has enormous potential for the production of all types of horticultural crops like fruits, vegetables, ornamentals, medicinal plants, spices and plantation crops in different parts of the country. Horticultural crops are known as a functional food and providing of healthy foods, vitamins and minerals. The horticulture sector is contributing more than 30 per cent to the agricultural gross domestic product (GDP) with a sustained growth rate of more than 5% in the last two decades. The current production estimate of horticultural produce is 341.63 million tonnes during year 2021-22 which surpassed the food grain production (315.72 MT). The per capita availability of fruits and vegetables in India is 207.9 g and 388.7 g per day, respectively. While, demand of fruits and vegetables is increasing day by day and estimated demand is 600 MT by 2050. Although, agriculture production is in upward trend, but still our food and nutritional security are under pressure due to certain factors like increase in population, inflation, income disparity and climate uncertainty. Simultaneously, emission of greenhouse gases and their alarming consequences imply the need for carbon sequestration mechanisms while the carbon credit mechanism urges for a green economy. On the other hand, India has only 24.62 % forest cover and Tree Outside Forests (TOPs), which is far below the recommended 33% of our National Forest Policy, 1988. Increasing the area under natural forests has negligible scope, while Trees Outside Forests (TOFs) offer tremendous opportunities for productivity augmentation and sustainability.

Opportunity and Present Scenario

At this juncture, perennial horticulture crops offer opportunities for climate change mitigation and food and nutritional security with additional advantages. There is an urgent need to increase the production of quality planting material (QPM) for perennial horticulture crops to combat various challenges through area expansion under perennial horticulture crops. This presents a challenge in front of the nursery sector to ensure the availability of quality planting material for area expansion. Generally, farmers do not have access to disease-free planting material for true-to-type varieties. With the existing infrastructure, only 30-40% demand for quality planting material is being met by the various National Horticultural Board (NHB) accredited nurseries established in Indian Council Agricultural Research (ICAR) Institutes, State Agricultural Universities (SAUs), line department, Centre of Excellences and in the private sector.

Opportunity in the Nursery Sector

There is a high and sustained demand for quality planting material in the horticulture sector and increasing over the years. This is providing opportunities for the establishment of new nurseries, entrepreneurship development, creation of jobs directly as nursery managers, skilled young professionals, propagators, and field supervisors, as well as indirectly in transportation, marketing and in the input supply chain. The nursery sector is a viable business model for educated youth and women through the formation of Farmers Producers Organization and Self Help Groups by taking benefits from different government schemes. Nurseries can play a major role in the overall development of the horticultural sector and in the national economy by exporting quality planting material to other countries. The availability of quality seedlings at lower cost offers ample scope for large-scale planting. A recognized Model Nursery should function as a reliable source of quality stock for horticultural crops. The success of a nursery depends upon the genetic purity of the mother plants. A good mother plant should be true to type, vigorous, high yielding, healthy, and should be free from insect pests and diseases. Nurseries must maintain a seed block, a scion block, and a stock bed in order to produce certified nursery stock and should use propagation material from these sources only. Management of horticultural nurseries involves several operations including grafting, budding, potting, repotting, transplanting, etc. Additionally, technological advances such as monitoring, disease detection, Robotics, automation etc. may help nurseries to become more efficient and competitive. The technical specification to produce horticulture plants made by ICAR is also helping in the production of quality planting material.

Issues of Horticulture Nursery Sector

The horticulture nursery sector is facing several challenges, including a lack of credit support, modern infrastructure facilities and mechanization, implementation of quality control and domestic quarantine, shortage of skilled labour, and poor in maintaining mother blocks thus, laggards in supplying improved varieties on desired rootstocks, and most of the nurseries small, family-owned business struggling for resources. Further, the nurseries located in the vicinity of cities are retail type and engaged in selling of sub-standard horticultural plants of unknown pedigrees which are serious concerns that need to be addressed. The emergence of new pests and diseases under changing climatic scenarios is also a challenge in the nursery sector.

In addition, there are other certain issues in the nursery sector like availability of land, cost of land, availability of quality water, maintenance of disease-free mother plants, and the emergence of new pests and diseases under changing climatic scenarios pose a hindrance to the growth of the nursery sector, needs the attention of policymakers and researchers. In India, the Nursery Registration Act has been in operation since 1966 and different state governments have also enacted their own nursery registration acts. However, many state governments still lack a legal system for the regulation of nurseries. One major issue is the lack of a quick and formal system of quality assurance regarding the varietal genuineness of planting materials needs to be addressed. India can emerge as a far bigger producer and exporter if the sufficient emphasis is given to resource allocation, infrastructure development, more R&D, technological upgradation and a better policy framework for horticulture nurseries as well as address the following issues properly.

- Availability of land, cost of land, availability of quality water.
- Maintenance of disease-free mother plants.
- The emergence of new pests and diseases under changing climatic scenario.
- The attention of policy makers and researchers.

Meeting the Challenges for Nursery Owners and Workers

To combat the immediate labour crisis, many nurserymen are willing to pay workers more, hence compel to increase the cost of the saplings, hence labour costs are much higher compared overall cost of nursery plant production. Furthermore, labour cost also influences by season, therefore the focus must be given to increasing mechanisation to reduce labour dependency and provision must be given to link nursery activities with MGNREGA.

Legal Authorizations for Starting Nursery Business

A commercial nursery business requires a business license which is a rather lengthy process. Normally, a nursery business would be considered an "agricultural use", but in some cases, it may also be interpreted as commercial", "agribusiness", or some other classification. Permits regarding the establishment of greenhouses, storage buildings or warehouses are required from the competent authority, which may be a lengthy process or may take longer time. Additionally, a nursery business may require property hazard insurance, workmen's compensation if you have paid employees and general liability insurance if you expect visitors to your nursery. Furthermore, nurseries in some areas may have to fulfil the government agricultural inspection requirements as per the body governing the area. Permission is also required from the local authorities for availing electricity, water, and other facilities.

Technological Advances in the Management of Hi-Tech Nurseries

Several innovations have been made in nursery management, ranging from compact tools like CO_2 analyzers, lux metres, thermostats, oxygen sensors, CO_2 generators, soil moisture sensors, dehumidifiers, and quality measuring devices to complete automation and robotics.

Micro-Irrigation: In nurseries, individual plants can be irrigated with low-pressure irrigation systems using micro emitters, bubbles and foggers can be used for misting, micro-sprinkle for cooling.

Fertigation: Fertilizers can be injected into drip irrigation systems by selecting appropriate equipment like valves, tanks, venturies and aspirators. Pressure Differential Injection System is the most common and works without a power supply by taking advantage of the system's pressure-head differences.

Weed control: Use of woven weed mates for covering the floor of the beds and paths which control weeds and permit percolation of rainwater.

Automation: Automating isn't simply buying a robot and writing a check, "It's planning and developing an entire system around the specific operation. In general, environmental control system used in ordinary greenhouses are a combination of timers, human adjustments, and theoretically regulated activities. However, modern computerized environmental control systems enable the integration of the many greenhouse components into a productive and efficient system. Automatic environmental control systems can maintain favourable environmental conditions. Automation of irrigation systems, which saves labour, enables accuracy in adjusting irrigation timing and volume. In a similar way, controls have made it possible to adjust the settings for heaters, ventilation fans, and wet pads. Through the provision of consistent, comprehensive data on the greenhouse environment, computerized control systems can aid in the development of a grower's overall management strategy.

Smart Crop Temperature Sensor

It is an infrared thermometer sensor used to measure plant canopy temperature. Every minute, the sensor measures the canopy's temperature, reporting an average measurement every 15 minutes. This system alerts gardeners via a text message or computer report when plants are thirsty or too hot for optimal growth.

Multi-Rotor Systems

A low-altitude multi-rotor system (MRS) is used as a low-cost platform to collect images for real-time inventory data. The MRS can fly over both container and field-grown plants to perform inventory tasks. It will offer a low-cost way to check weed or insect pressure, or monitor plant stress as needed. It will process images of thousands of plants to actual counts of certain grades of a specific variety. These details might be connected to a sales prediction. Growers might potentially know the precise location of every plant and details about each plant because photographs are GPS-tagged (Robbins et al, 2011, University of Arkansas, jrobbins@uaex.edu).

Use of Robotics

The companies like Harvest Automation have developed container-moving robots. These robots can combine, relocate, or space out plants. Each machine has a 4 hour battery life and can move 200–300 containers each hour.

Sales App and Seed Portals

As more and more mobile apps were released over the past few years, the phrase "There's an app for that" started to sound like a rallying cry. An iPhone and Android mobile app named MofuMobi was just made available by SBI Software. Users of this app get immediate access to vital sales and marketing documents, including sale reports, contact details for clients, and weekly and monthly sales numbers. Users of the app can access up-to-date figures while on the go. ICAR - IIHR has developed Seed Portal which helps the farmer to know real-time availability of plants, and prices and is linked with the delivery system.

Hi-tech nurseries improve physiological and physical processes that help plants function well. They also increase seed germination rates, germination percentages, and grafting success, lengthen the propagation period, and make it possible to successfully propagate plants in extremely harsh and unfavourable climatic conditions. A few nurseries of the private sector have successfully developed a reputation in the market. They are producing QPM on large scale (tissue cultured and grafted plants) by use of the advanced technique of plant propagation and modern infrastructure like Hi-Tech poly houses, mist chambers, mechanization, disease monitoring& detection, efficient nursery tools and gadgets, and implements.

Initiatives for Horticulture Nursery

A series of initiatives are being taken to provide assistance to the public and private sectors for setting up new Hi-Tech and small nurseries (minimum area should be 1 ha), by governments credit push through Agri Infra fund (MoA & FW), project financing by NHB, NABARD, Cooperative banks, Small finance banks, RRBs etc. for the development of the nursery sector. Mission for Integrated Development for Horticulture (MIDH), a centrally sponsored scheme, promoting quality plant material production through the setting up and modernization of nurseries, and strengthening of the tissue culture units. Under MIDH guidelines for establishing Hi-Tech Nursery (4ha) and a small nursery (1 ha), cost norms are Rs. 25 lakh/ha and Rs. 15 lakh/ha and 100% public sector respectively (https://midh.gov.in/). National Horticulture Board (NHB) has developed guidelines for licensing, accreditation; star rating of horticultural nurseries and it is based on rating on a scale of single star to five star with appropriate weightage on the production system, nursery management practices, and quality of planting material produced (https://www.mofpi.gov.in/). Recently, NHB has launched National Nursery Portal that provide interface to sellers and buyers. On this buyers could access real time availability, prices of nursery plants and status of registration and rating of the nurseries. Presently, there are 741 NHB accredited and star rated nurseries. A big initiative had been taken under the Indo-Israel mission, which led to the development of thirty-four Center of Excellence (CoE) during 2008 to 2022 on various fruits and horticultural crops. These are involved in the development of production technologies, the production of high quality planting material, and also involved in knowledge transfer and skill development.

- Project financing: Cooperative banks, Small finance banks, NCDC, NBFC, NABARD, APEDA, RRBs.

- NHB: Guidelines for licensing, accreditation, star rating of nurseries, National Nursery Portal, Cluster development programme, subsidised funding under MIDH.
- MoA&FW: Credit push through Agri Infra Fund.
- Indo-Israel mission: 34 Center of Excellence (CoE).
- Formation and promotion of FPOs, SHG.
- Engaging MNREGA laborers with nurseries.
- Mandatory procurement of QPM for MIDH/Govt. subsidised schemes from accredited nurseries.

Seed Portals and Sales Apps

- **Seed Portal-** by ICAR - IIHR to know real time availability of plants, prices and linked with delivery system.
- Mobile apps
- **MofuMobi** made available by SBI Software (iOS and Android) for sale management.

National Nursery Portal by NHB

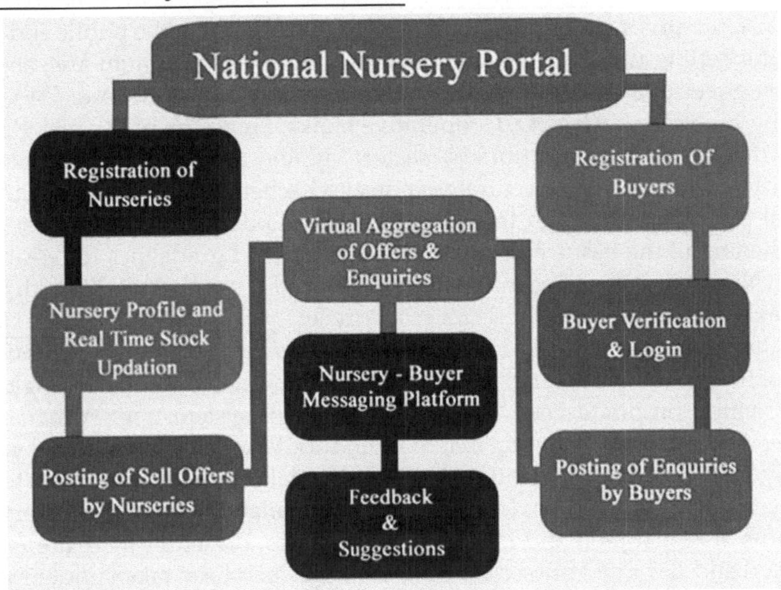

Path Ahead

- Need to apply molecular markers in variety identification in the nursery
- Phytosanitary and bud wood certification should be made mandatory
- Application of hyperspectral imaging (HSI) based modelling of plant varieties for testing of genuineness of varieties.

- Ensure domestic quarantine mandatory
- Umbrella supervision for all small nurseries to insure the safe production of saplings
- Regulated outlet

References

Burton, L. (2017). Climate change: risks and opportunities in nursery production. *Acta Horticulture*. 1174, 5-12.

Haase, D. L. (2010). Strategies and challenges for nursery production: Perspectives on where we're going and where we've been. *National proceedings: forest and conservation nursery associations-2009. Proc. RMRS-P-62. US Department of Agriculture, Forest Service, Rocky Mountain Research Station, Fort Collins, CO*, 38-44.

Harrison, S., Gregorio, N., & Herbohn, J. (2008). A critical overview of forestry seedling production policies and practices in relation to smallholder forestry in developing countries. *Small-scale Forestry*, *7*, 207-223.

Jain, V. L. D. B. (2022). Constraints faced by plant nursery owners in Raipur, Chhattisgarh. The Pharma Innovation, 11(6),1408-1410.

NHB (2022). Indian Horticulture Database. Gurugram, India. http://www.nhb.gov.in.

Sengar, R. S., & Rani, V. (2020). Opportunities and prospective of integrated development of horticulture: A review. *Annals of Horticulture*, 13 (1), 1-8.

Xu, X. (2022). Major challenges facing the commercial horticulture. *Frontiers in Horticulture*, 2.

Robbins, J., Owen, J., Saraswat, D., Ehsani, R., Maja, J., Stoven, H., Landgren, C. and Doane, S., (2011). Development of an inventory management tool using visual imagery from a multirotor aerial platform. In *Proc. of the Southern Nursery Association Annual Research Conf* (Vol. 57, pp. 8-12).

Chapter 5

Exotic Fruit Cultivation in India Potential and Needs

Vishal Nath, Krishna Prakash and Swapnil Pandey[1]
ICAR- Indian Agricultural Research Institute,
Gauria Karma, Hazaribag, Jharkhand
[1]*Fruit Research Station (PAU), Ballowal, SBS Nagar, Punjab*

Although India is a rich heritage of indigenous tropical, subtropical and temperate fruits, many exotic fruit species have created space in Indian subcontinent and being grown commercially in many parts of the country. The term 'Exotic fruits' denote those fruits which are not native to an area but are being cultivated and are made out of their place of origin. Some such fruits have been well merged and looks like native however some are picking at faster pace. Fruits like kiwifruit, dragon fruit, rambutan, longan, mangosteen, passion fruit, loquat, parsimon, pulasan, durian, avocado and many other fruit species are being popularized in different regions of the country whereas, fruits like litchi, cashewnut, apple and many others have set well in the region. With the increase in purchasing capacity of customers in India, and the growing health consciousness among people, the demand for non-traditional fruits has been increased in the country over the last couple of years. And thus, the fruit markets in India has been thriving up with sizeable share of diverse fruit products of exotic fruit crops. Interestingly, the demand and supply of exotic fruits is penetrated down the local fruit markets in remote villages, partly due to sporadic cultivation in pockets and wholly due to ease of import from the other countries. Selling of different coloured dragon fruits, hairy rambutan, creamy avocados, mangosteen, and many more variable exotic fruits are not a rare sight today at the local fruit vendors. These exotic fruits have immense potential to contribute location specific food production, as they are well adapted to existing and adverse environmental conditions. A lot of enterprising farmers are taking up the cultivation of exotic fruits in their backyard or well organised orchards to meet the new and growing demand and enable the consumers to relish the health benefits of exotic fruits and their products.

The agro-climatic condition of India offers an excellent scope for growing of different types of exotic fruits within the country (Fruits like kiwi, rambutan, mangosteen, avocado etc.) or the whole country (Fruits like dragon fruit, longan etc.). Anticipated increase in production of exotic fruits will not only reduce the dependency of mass livelihood improvement and reduced burden of nutritional security on few fruits but also help in diversification of agriculture-horticulture besides catering multifarious human needs. These crops may also be providing many fold employment opportunities and agro based industries, packaging, storage, preservation, canning and product diversification. Therefore, expansion of area under exotic fruits through adoption of improved package of practices for cultivation is required to increase the fruit production in the country.

Why Think about Exotic Fruit Crops?

India is second largest fruit producer in the world. However, only a few fruits are grown commercially, while most of the others are given less importance or not traded widely and rendered them as underutilized or underexploited or minor fruit crops. Most of the exotic fruits comes under the later category of fruits which have been introduced to our country in due course of times however, some of these have covered a considerable area of cultivation and few are still confined in backyard of an innovative farmer or small portion of a fruit orchard. Sometimes, growing exotic fruits is fascinated also but after its adoption and realization of potential, it is good too. Litchi, if we consider as an exotic crop, which was introduced in India during 18th century, was grown as privileged fancy crop today Indianised and occupies an area of more than one lakh hectare and production curve is rising with expansion of growing areas. Similarly, Dragon fruit was introduced in India during early 90's and occupies an area of 3000 hectares today while government is planning to increase the area under dragon fruit upto 50,000 hectares in next 5-10 years. One of the major advantage with India is that the country has substantial area under marginal and wastelands of different kinds viz. ravine, hilly tracts, alkaline, acidic, marshy, occasional flood, sandy soils etc. which are not suitable for cultivation of high input demanding horticultural crops, either due to poor quality of soil or lack of input resources. Such lands may be utilized for growing of low input demanding potential underexploited or exotic fruits (As in case of Dragon fruit, the Kamalam). The unique tastes, higher profit and localized importance help in region specific large scale cultivation of some of the fruit crops (As in case of Litchi). Sometimes, there has been growing consumer interest in "exotic" terminology, may use to be referred as 'mysteriously different' or 'unusual' fruits due to anecdotal claims of their medicinal values and purported health benefits. And thus there will be a great demand in market (As in case of Kiwi, Passion fruit, Rambutan, Longan, several Annona sp. etc.).

Exotic fruits provide an opportunity to diversify the present day fruit availability, which is so inevitable in views of increasing population pressure and fast depletion of natural resources as well as changing human needs. Many of the exotic fruits and their products are very rich sources of vitamins, minerals, antioxidants and other nutrients besides major nutrient components of carbohydrates, proteins and fats. A lot of consumers are adopting plant based foods and incorporating nutritious exotic fruits in their daily food habits due to their wonderful flavour and myriad health benefits. For example, avocados are a great source of Vitamins C, E, K and B-6, the sweet and tangy kiwis are loaded with nutrients like Vitamins C, K, E, folate and potassium. Many of the exotic fruits have been known for their medicinal and therapeutic properties and are promoted to cure various ailments through nutritive support. Huge amounts of funds are invested for scientific research on exotic fruits and their health benefits in our country.

Quite a few of these exotic fruits have excellent flavour, aroma, attractive colour and are suitable for preparing several processed products. These fruits have unlimited potential in the processed form and consumers all over the world can get the opportunity to enjoy the fruits in the form of their products. A wide array of exotic fruits that appear in the supermarket aisles and on the rickety carts of the street hawkers are reminders of the changing scenario in the country, now, probably due to the interest in man's quest for something different than the traditional fruits. Many exotic fruits may bring a refreshing change in our platter which was getting bored of having the same fruits taste year after year. Moreover, exotic fruits fill up the shops during off-season of a fruit crop also. For example, local apples are generally available during the winter season, while imported apples are seen available round the year and are eventually becoming more popular.

Moreover, climate change threatens to destabilize production and, as the global population shoots up, the green revolution is reaching its limits in generating the ever-increasing amounts of food needed to feed the people of country. Bringing exotic crops into the mainstream cultivation orchards may reduce the unforeseen risks associated with continuous cultivation of major fruit crops in an area. Many of the exotic fruit crops usually thrive in infertile or difficult terrains that are not well suited to large scale commercial agriculture, have exotic fruits can be a viable option.

Scope of Exotic Fruit in India

The consumers in India are becoming more health-conscious and are searching for satiating options for their nutritional needs. This is the driving force behind witnessing the growth of exotic fruit business in India due to increase in imports as well as attempts of some local farmers towards growing fancy home-grown produce.

India's annual import of exotic fruits is gradually growing over the years. Fresh fruit imports to India are pegged at more than 7 lakh tonnes in 2022 and a major portion of it was occupied by so called 'exotic' fruit. Roughly exotic fruit market which has Rs. 10,000 crore size business also brings with it a plethora of opportunities. The business of exotic fruits is largely influenced by the ever-evolving consumer habits and the abundant health benefits they carry. The demand for exotic fruits is not only witnessed in metro cities, but also in tier II, III cities as well as small towns. It is also noted that the exotic fruits and products is normally priced higher and commands a premium of approx. 50 percent more over local ones. For example, imported avocados are priced between INR 200 and INR 400 per piece, compared with local ones at 100-150 INR and similar is case of Dragon fruit and many more. It is needless to mention that, the farmers of exporting countries of these commodities, create a special attraction for these fruits in terms of size, shape, colour, or even taste for that matter to attract more and more consumers and fetch better price.

In India also, farmers from Northern Kashmir to southern states like Kerala and Karnataka are showing interest in the cultivation of exotic varieties and fruit crops to earn good returns. In fact, farmers of these regions are directly supplying their produce to restaurants, enabling chefs to whip up exciting dishes every day. Thanks to more affluent and well-travelled Indian consumers that the importers and retailers of the country are witnessing a surge in demand for the exotic fruits and fruit products.

Fruits like dragon fruit – a superfood that is almost entirely imported from South-east Asia, is witnessing a sharp increase in demand in India in last 10 years. Its potential has led many farmers in the many Indian states to cultivate the crop locally to meet the rising demand to what they believe would be the next big lucrative produce. Government is also promoting this crop in many states. New name 'Kamalam' has been given in India for Dragon fruit to bring it more closer to Indian minds and hearts.

Exotic Fruits Grown in India

Though, many exotic fruits are grown/cultivated in India at various scales however, some of the potential exotic fruit available in India has been listed in Table.

Table 1: Potential exotic fruit crops of India

A) Well versed Exotic Fruits in India

Common name	Year of Introduction to India	Scientific name	Family	Flowering	Fruiting
Litchi	18th Century	Litchi chinensis	Sapindaceae	Dec-Jan	May-July
Guava	Early 17th Century	Psidium guajava	Myrtaceae	Feb-March June-July October	July-September November-Jan February-April
Cashewnut	16th Century	Anacardium occdentale	Anacardiaceae	October-November	February-May
Apple	1865	Malus domestica	Rosaceae	Feruary-May	Sept-October
Pineapple	1548	Annanas comosus	Bromeliaceae	-	May-August
Papaya	Late 16th Century	Carica papaya	Caricaceae	Round the year	Round the year
Sapota	Late 18th Century	Manilkara achras	Sapotaceae	Round the year	Round the year
Cacao	1798	Theobroma cocoa	Sterculiaceae	Round the year	Round the year

B) Upcoming Exotic Fruits in India

Common name	Scientific name	Family	Flowering		
Kiwifruit	Actinidia deliciosa	Actinidaceae	February-March	Oct-Nov	Fresh, beverages, culinary
Dragon fruit	Pitaya hylocereus	Cactaceae	May-June	Sept-Oct	Juice, wine
Passion fruit	Passiflora edulis	Passifloraceae	April-May	June-Aug	RTS, beverages
Avocado	Perseaamericana	Lauraceae	Febuary-March	July-Aug	Fresh, shakes
Carambola	Averrhoea carambola L.	Oxalidaceae	June-July	Dec-Jan	Eaten fresh, RTS
Rambutan	Nephelium lappaceum	Sapindaceae	March-April June-July	June-Oct	Fresh, RTS
Longan	Dimocarpus longan	Sapindaceae	March-April	July-Aug	Fresh, resins
Pulasan	Nephelium mutabile	Sapindaceae	March-May	Aug-Octobr	Fresh, RTS
Fig	Ficus carica	Moraceae	January-March	April-June	Fresh, dried products
Loquat	Eriobotrya japonica	Rosaceae	October-November	March-April	Fresh fruit, jam, jelly, squash
Persimon	Diospyros kaki	Ebenaceae	Feb-March	October-November	Fresh, dried product
Mangosteen	Garcinia mangostana	Clusiaceae	April-June	October-November	Fresh, processed
Elephant apple	Dillenia indica	Dilleniaceae	June-July	October-November	Chutney, pickle
Tree tomato	Cyphomandra betacea	Solanaceae	Round the year	Round the year	Fresh, chutney
Durian	Durio zibethinus	Bombacaceae	July-August	October-November	Fresh, processed

Brief information about some of the upcoming, nutritious exotic fruits which are being grown in India have been mentioned here.

1. Kiwifruit (*Actinidia deliciosa*)

Kiwifruit or Chinese gooseberry is known as 'China's miracle fruit, and 'the horticultural wonder of New Zealand', a recent introduction in India. From China it is spread to New Zealand where it was recognized as a potential fruit and became a popular backyard vine. In India, the area under this fruit is negligible, being a new exotic introduction. In India, it is grown in mid hills of Himachal Pradesh, Jammu and Kashmir, Arunachal Pradesh, Meghalaya and Kerala. Kiwi can be grown in areas experiencing 700-800 chilling hours (No. of hours during which temperature remains at or below 7^0 C during the winter season). Fruits are rich in minerals, sugars, vitamins and carbohydrates. Besides, it has a refreshing delicate flavour with pleasing aroma. Fruits can be eaten fresh or processed in to squash, juice and wine. At present seven cultivars of kiwi namely Allison, Monty, Hayward, Abbott, Bruno (Female) and Tomuri and Matua (Male) have been introduced in India. Planting distance varies according to system of training. January is ideal time for planting of kiwifruit. Kiwi is dioecious plant (means male and female flower are born on different plant) thus needing plants of both sexes to produce fruit. A staminate plant is provided for the pollination of every six-pistilate plant. Every third plant in alternate row should be a pollinizer (staminate). Kiwi plants are pollinated mainly by honeybees besides hand pollination in early morning hours are very effective for heavy fruit setting and quality yield. It is commercially propagated by cutting and tongue grafting during dormant season (Dec.-Jan.). Training is required to establish and maintain a well-formed framework of main branches and fruiting arms. Vines are trained in two arm kniffin system. Bearing starts at the age of 4-5 years. Yields vary from 25-100 kg /vine (Patel *et al.*, 2008a). It ripens from October to December depending upon variety and climate. So far, no serious pest and diseases attack has been reported in this fruit from India. Therefore, it has got a better scope to become commercial eco-friendly fruit crop in foot and mid hills of the region.

2. Dragon Fruit (*Pitaya hylocereus*)

Dragon fruit is a tropical fruit that has become increasingly popular in recent years primarily due to unique look and taste. It is a cactus fruit that can be found throughout Asia, Australasia, North America and South America, even though they are believed to be native to Mexico originally. This fruit belongs to Cactaceae family also known as strawberry pear or pitaya. The fruit plant can thrive well in poor soils and temperature variations. Commercial cultivation of dragon fruit is picking up in India due to high market price. There are two main types of dragon, the sour types, typically eaten in the Americas, and sweet types, found across Asia. The fruit comes in 3 different color varieties, labelled as red, yellow and Costa Rican dragon. The "red" fruits are generally a bright magenta color on the outside, with yellow flesh. The Yellow dragon fruit is yellow inside and out, and the Costa Rican dragon are magenta on the outside and the inside. Dragon fruit contains small amounts of several nutrients. It's also a decent source of iron, magnesium, and fiber. They smell deliciously fragrant and most have a sweet flavor similar to a kiwi fruit. Studies suggest that dragon fruit may provide various health benefits likely due to its high fiber and antioxidant content. It is low-calorie fruit that contains less sugar and fewer carbs than many other tropical fruits. Juice or wine can be obtained from the fruit, while the flowers can be eaten or used for tea.

3. Passion Fruit (*Passiflora* spp.)

Passion fruit is a high value export-oriented crop owing to its excellent flavoured juice. This is a native of Brazil and widely cultivated in South Africa, Kenya, East Africa, Australia, New Zealand and Indonesia. In India, it grows wildly in Niligiri Hills, Kodaikanal, Coorg, Malabar, Kerala, and Himachal Pradesh. It is commonly cultivated in warm and tropical regions. Recently its cultivation has been extended to some areas of north eastern hill region (Patel *et al.*, 2008b). The juice is a good source

of vitamin 'A' and 'C', and used to flavour ice cream, making jam and refreshing drinks. Some home-scale products like squash and nectar can be prepared from its juice. Further, the excellent flavour of juice can be utilized to enhance the overall quality of the blended products. Its cultivation is extended to South India and NE region. Passion fruit is commercially propagated by cuttings and seed. The passion fruit (*Passiflora* spp.) belongs to the genus Passiflora, which includes about 500 species. There are at least ten species of the genus *Passiflora* that are cultivated for their fruits. Among these species, only purple passion fruits (*Passiflora edulis* Sims) and yellow passion fruits (*Passiflora edulis f. Flavicarpa* Degener) are commonly cultivated. The yellow passion fruit is slightly larger than the purple form and has a bright canary-yellow rind. The pulp is somewhat more acid and the seeds are dark brown rather than black. Yellow type species is prone to frost. Therefore, initially it takes more time at least two years to become hardy and give the satisfactory yield. Commercially Purple, Yellow type and Hybrid (Kaveri) cultivars are being cultivated in India. It is very vigorous, high yielding, tolerant to collar rot, wilt, brown leaf spot and nematodes. Its fruits are a blend of purple and yellow varieties but having the size of the yellow variety. The pulp is less acid than that of the yellow variety.

4. Avocado (*Persea americana*)

Avocado (Butter fruit) is a native of tropical America. It originated in Mexico and Central America, possibly from more than one wild species. The early Spanish explorers recorded its cultivation from Mexico to Peru but it was not in the West Indies at that time. In India, avocado is not a commercial fruit crop. It was introduced from Sri Lanka in the early part of the twentieth century. In a very limited scale and in a scattered way it is grown in Tamil Nadu, Kerala, Maharashtra, Karnataka in the south-central India and in the eastern Himalayan state of Sikkim. Avocado is the most nutritive among fruits and is regarded as the most important contribution of the New World to human diet. The fruit is relished by some people, but not by others. The pulp is rich in proteins (up to 4%) and fat (up to 30%), but low in carbohydrates. The fat is similar to olive oil in composition and is widely used in the preparation of cosmetics. Avocados have the highest energy value (245 cal/100 g) of any fruit besides being a reservoir of several vitamins and minerals. Avocado is mainly used fresh, in sandwich filling or in salads. It can also be used in ice creams and milk shakes and the pulp may be preserved by freezing. Climatically, it is grown in tropical or semitropical areas experiencing some rainfall in summer, and in humid, subtropical summer rainfall areas. It can not tolerate the hot dry winds and frosts of northern India. It can successfully grow under water logged condition. The cultivars of West Indian race are grown in localized pockets in Maharashtra, Tamil Nadu and Karnataka.

In tropical and near-tropical areas, only West Indian race is well-adapted but its hybrids with Guatemalan (e.g. Booth selection) perform well and are considered valuable for extending the harvest season. In the eastern Himalayan state of Sikkim, avocado has been introduced successfully in hill ranges with an altitude of 800 to 1,600 msl. Both the Mexican and Gautemalan races are grown successfully in Sikkim. The varieties that are cultivated in India go by several names, such as Purple, Green, Fuerte, Pollock, Peradeniya Purple Hybrid, Trapp, Round and Long. Among the several existing varieties, perhaps Fuerte is the most widely grown, but it is regarded as unsuitable for the tropics Avocado is commonly propagated through seeds. Avocado is planted out to a distance of 6 to 12 m depending on the vigour of variety and its growth habit. Among insect pests, scales, mealy bugs and mites are the important ones, and may be controlled by suitable insecticides. The yield ranges from about 100 to 500 fruits per tree. In Sikkim, specifically, on average 300-400 fruits can be harvested from 10-15 years old trees. In Sikkim, fruits of Purple variety are harvested during July, and for Green variety September-October is the usual harvesting time. In Tamil Nadu, July-August is the peak harvest time. The yield performance of avocado, both in tropical southern India and humid sub-tropical north-eastern India is highly satisfactory.

5. Rambutan (*Nephelium lappaceum*)

Rambutan is a juicy tropical fruit belong to the family Sapindaceae and the aril resembling to litchi fruit. Its aril is sweeter and crisped than litchi or longan aril. The fruit is native to Indonesia and Malaysia. In India it is commercially cultivated in Kerala, Karnataka, T.N., NE states and has possibility of growing of cold tolerant variety in Bihar. Rambutan is evergreen tree growing to a height of 12–20 m. The fruit is a round to oval drupe 3–6 cm (rarely to 8 cm) tall and 3–4 cm broad, borne in a loose pendant cluster of 10–20 together. The leathery skin is reddish (rarely orange or yellow), and covered with fleshy pliable spines. Flowering periods differ for other localities. In south India, rambutan flowers from March to July and again between June and November. Rambutan require warm tropical climates, temperature around 22–30 C, and is sensitive lower than 10 C, The fruit matures 15–18 weeks after fruit set. The important cultivars grown in India are Arka Coorg Arun, Arka Coorg Patib, N-18 and E-35 (Jacob *et al.*, 2013). It is commercially propagated by grafting, air-layering and budding. Budded trees may fruit after 2–3 years with optimum production occurring after 8–10 years. Trees grown from seed bear after 5–6 years. Planting distance should be kept 5-6 m apart and planted during monsoon season with provision of pollenizer. Training and pruning should be done to give a strong frame work and for managing the plant canopy after harvesting of the fruit. Rambutan is infested by leaf-eating insects, mealy bug, (*Pseudococcus lilacinus)* and giant bug (*Tessaratoma longicorne)*. Powdery mildew, (*Oidium* sp.) and stem canker (*Fomes lignosus)* are major disease of concern.

An average 60–70 kg fruit per tree fruit yield may be obtained regularly. The ripened fruits continue to be staying on the trees up to 30 days. Fruits have high demand in Europe, the Middle East and North America market as fresh fruits. The Rambutan fruit is rich in nutritive value and has very rich source of energy. It is rich in minerals and vitamins.Seeds are light brown, high in certain fats and oils (primarily oleic acid and arachidic acid) and valuable to industry, used in cooking and the manufacture of soap. Rambutan roots, bark and leaves have various uses in medicine and in the production of dyes.

6. Longan (*Dimocarpus longan* Lour.)

The longan is native to southern China and hills of Assam and the Garo hills. It can be grown between elevations of 150-450 msl. It is a sub-tropical prolific bearer fruit also known as dragon eye fruit. The tree is erect 9-12m in height and 14m in width, spreading, slightly drooping and heavily foliaged branches. It requires a cool climate for its growth, particularly for flowering. The longan prefers soil pH 6.0 to 6.5 and cannot tolerate acidic soil condition. In India, West Bengal is a prominent longan growing state. Longan season in Southeast Asia is mid to late summer, from July through September. There is an off-season period in Thailand from November to early February as well. . Most longan trees have been grown from seed. It is planted at 5-6 m apart during monsoon season. Longan is normally propagated by air layering. The longan is relatively free of pests and diseases.

The fruits, in drooping clusters, are globose 1.25-2.5 cm in diameter, with thin, brittle, yellow-brown to light reddish-brown rind, more or less rough, the protuberances much less prominent than litchi. Longans are also renowned in Indian and Chinese cultures for their blood-boosting benefits: the fruits stave off anaemia and boost iron absorption. Furthermore, the high antioxidants improve skin health and keep signs of aging at bay. Perfectly ripe longans are sweet, refreshing and clean with hints of floral nectar. It has 60 kcal energy, 15 g carbohydrate, 84 mg vitamin C, 10 mg magnesium, 21 mg phosphorous, and 266 mg potassium in 100g of edible longans. At times, there may be signs of mineral deficiency which can be readily corrected by supplying minor elements in the fertilization program.

7. Pulasan (*Nephelium mutabile*)

Pulasan is a fruit from South East Asia and is native of Myanmar, Indonesia, Malaysia, Thailand and the Philippines. It is mostly grown in lowland primary forests, often on river banks but rarely in swamps, and usually on sand or clay soils. In India it is grown Kerala, Tamil Nadu and some parts of Karnataka. An evergreen dioecious tree reaching 9-14 m, with a short trunk and a wide, rounded crown. The leaves and roots are employed in poultices. Root decoction is administered as a febrifuge and vermifuge. It is commonly propagated by air-layering and grafting (Varghees and Lukose, 2013). The pulasan is ultra-tropical and thrives well only in very humid regions between 110-350 m of altitude. In Malaya, it is said that the tree bears best after a long, dry season. The trees require less space than rambutan trees and can be planted at 8 to 10 m apart each way. 'Asmerah Tjoplok, Kapoelasanmeratjoplok, P-15 and Seebabat are important cultivars. As a rule, they receive little or no fertilizer or other cultural attention.

Pulasan fruits are harvested twice in a year. The season of trees flowering is from March-May and August-October. After flowering, the fruits start to mature in 15-18 weeks time. It takes 100-130 days to mature, in warm tropical areas. The first season of fruiting is from July-November and the next one is from March-July. Fruit is harvested in bunches when most of them turn yellow or red color. The grafted trees produce fruits in 4-5 years. The fruit contains per 100 g edible portion: water 85 g, protein 0.8 g, fat 0.6 g, carbohydrates 13 g, fibre 0.1 g and ash 0.4 g, vitamin C 10.8 mg. The dried seed kernels yield mildly perfumed oil. This oil solidifies into a white fat when the temperature falls below 105 F. This oil can be used for soap making. The wood is light red, harder and heavier than that of the rambutan and of excellent quality but rarely available.

8. Fig (*Ficus carica*)

Fig was an important food crop in ancient civilization. It is an Asian species native to the Middle East and Western Asia. This plant is cultivated or grown throughout the world either as an ornamental plant or as a fruit bearing plant. This plant is deciduous in nature and has fragrant leaves. In India fig farming is mostly done in

Maharashtra, Gujarat, Uttar Pradesh, Karnataka and Tamil Nadu. Soil with well-draining property and good water holding capacity are ideal for the cultivation of fig, especially alluvial clay loam or medium black soils. This fig tree is a subtropical tree that can tolerate a lower temperature range of 9.5 to 12 C. Fig tree propagation is done by various techniques such as budding, hardwood cutting, air layering and grafting. However, the most commonly used methods for commercially propagating the trees is through hardwood cuttings. The fig tree produces fruits after 2 years of planting. The plants produce a commercial harvest only in the third year of planting. The yield from the trees increases every year until the eight years and then stabilizes up to 35 years. The main harvest season is during February –March and is almost over by the end of summer.

Generally the fruits are harvested manually with an interval of 2 or 3 days. Fresh fruits that are ripe can be sold in the local markets. It is a highly nutritious fruit consisting of 84% pulp and 16% skin. Besides, the fruit also contains protein, calcium, iron, vitamin A and thiamine at varying concentrations. The fig fruit is highly perishable in nature and is not suitable for transport over long distances. Figs are consumed as fresh or dried, preserved, candied or canned. Fresh figs are nutritious and used as dessert or for making jam, jelly, pudding, cakes, etc. The fruit is valued for its laxative property. It is applied for boils and other skin infections. The latex is used to coagulate milk and leaves are used medicinally as diuretic, demulcent, emollient and anthelmintic properties.

9. Loquat (*Eriobotrya japonica* Lindl.)

The loquat (Japanese plum) fruit is native to China belongs to family Rosaceae. Many good loquat cultivars also originated in China and Japan. It has become a commercial fruit of Australia, U.S.A. Hawai, France and India. U.P. occupies the first position as for as area under loquat is concerned. Loquat is highly preferred fruit for kitchen gardens in Punjab (Nath *et al.*, 2009). It is also being grown on a small scale in the orchards at Amritsar, Gurdaspur, Hoshiarpur and Patiala districts. It is also found promising in Delhi, Himachal Pradesh, Assam, Maharashtra, and a few of the hill stations in the south. Loquat has adapted well to the harsh climate of North India.

It is performing well in tropical climate and lower hills, where temperature rarely falls below 0 C and can be grown up to 1600 m asl in India. Where the climate is too cool or excessively warm and moist, the tree is grown as an ornamental but will not bear fruit. As subtropical fruit, loquats are grown most successfully in citrus-producing areas. Loquats are quite resistant to most diseases and insect damage. Loquat fruits should be allowed to ripen fully before harvesting. They reach maturity in about 90 days from full flower opening. The commercial cultivars of India are Golden Yellow, Improved Golden Yellow, Pale Yellow, Tanaka, California Advance and Improved Pale Yellow. It is propagated by grafting. Planting distance should be kept about 6-7 m apart.

The trees are evergreen, spreading and erect growing. A full grown tree may attain a height of 7-8 meters. Selected varieties produce clusters of excellent yellow fruit that mature in the month of March-April. It contains 60 to 70 percent pulp depending upon the cultivar and 15-20 percent seed. The pulp is composed of 88% water, 0.6% proteins, 9.6% carbohydrates, 0.5% minerals, 1.0% fibre and 55% IU vitamin 'A'. Some preparations like Jam, Jelly and squash are made from the pulp. It is consumed as a fresh fruit. The colour of epicarp usually ranges between yellow-red. The fruit tastes sweet and sour with good flavour.

10. Persimmon (*Diospyros kaki*)

Persimmon is a deciduous fruit tree which is predominantly cultivated in sub-tropical and warm temperate regions belong to family Ebenaceae which is the most widely cultivated species is the Asian persimmon. Ripe fruits of the cultivated strains are in light yellow-orange to dark red-orange depending on the species and variety. Persimmons are eaten fresh, dried, raw, or cooked. When eaten fresh they are usually eaten whole like an apple or cut into quarters. Compared to apples, persimmons have higher levels of dietary fiber, sodium, potassium, magnesium, calcium, iron and manganese, but lower levels of copper and zinc. The persimmon market in India is limited for a number of reasons, mostly because of low demand and paucity of supply. However, several of India's cooler regions grow persimmons viz., Jammu and Kashmir, Himachal Pradesh and the hills of Uttar Pradesh. The

temperate regions of Kullu, Mandi, Solan, Chamba, and Shimla of HP are best for persimmon cultivation. A few trees also grow in the chilly Coonoor hills of Tamil Nadu. It has a relatively low winter chilling requirement in comparison with other deciduous fruit crops. The tree tends to have a higher tolerance to heat, water stress and water logging than other deciduous fruit crops. Persimmon cultivation has potential, though, as a single tree may bear 200 kilograms of the fruit per year. Fruiting season begins in mid-September and lasts until November, sometimes December. Commercially and in general, there are two types of persimmon fruit: astringent and non-astringent.

11. Elephant Apple (*Dillenia indica* L.)

It is originated in Indonesia and this evergreen tropical tree belongs to family Dilleniaceae and also called as Chalta. It is grown in Meghalaya, Assam and other parts of the country. This is a spreading tree and has beautiful white fragrant flowers, toothed leaves and globose fruits with small brown seeds. Greenish-yellow fruit, which has a thick protective covering, is edible; unripe fruits are cooked to make pickle and chutney (Patel *et al.*, 2009). Juicy pulp is aromatic but very acidic.

12. Tree Tomato (*Cyphomandra betacea*)

It is native of South America belongs to Solanaceae family, a perennial shrub producing red tomato like fruit used for making chutney. It is grown as backyard venture crop in India especially in Meghalaya, Mizoram Manipur, Nagaland and some parts of South India (Patel *et al.*, 2013). It is a small, tender 2-3 m tall tree which bears prolifically egg shaped berries withpointed ends in cluster near the young shoots. The long-stalked, pendent fruit, borne singly,or in clusters of 3 to 12, is smooth, egg-shaped but pointed at both ends and capped withthe persistent conical calyx. In size it ranges from 5-10 cm long and 4-5 cm in width. Skincolor may be solid deep-purple, blood-red, orange or yellow, or red-and-yellow, and mayhave faint dark, longitudinal strips. The inside pulp of fruits is light orange and seed isblack. Tree tomato is consumed as chutney when raw or after roasting and peeling off theskin. It is liked by the people due to unique falvour. It is propagated by both seed and cutting.

13. Durian (*Durio zibethinus*)

Durian is a native to Borneo belongs to family Bombacaceae, produce fruits with a unique appearance, taste, flavour and aroma. The durian is grown intensively and commercially for its fruit only in Indonesia, Malaysia and Thailand. In India, there are no large orchards or commercial plantings of durian, but for some trees in and around Nilgiris (Tamil Nadu) and West Coast. This fruit tree grows tall and straight to a height of 30 m in the forests. However, grafted orchard trees seldom grow over 12 m. The durian fruits are ovoid to ellipsoid, large, weighing up to 5 kg with thorny surface like jack and normally olive green in color. People have strong like or dislike for this fruit. Many like it, because of its sweet, delicious and filling

taste, while others dislike it for its highly objectionable odour. The penetrating odour is comparable to that of rotten onion and is unacceptable to many people. However, once the odour barrier is overcome by a strong determination, the experience of eating a good durian is never forgotten. The pulp is rich in sugars (12%), protein (2.8%) and carbohydrates (34%) in addition to iron, B vitamins especially the uncommon but valuable vitamin E. When durian fruit is consumed, it gives a feeling of internal warmth, followed by a glowing sensation, and this has led to a strong belief that the fruit has aphrodisiacal qualities. The fruits are highly prized both as fresh fruit and in processed form. Durian also has some pharmacological properties, e.g. decoction of roots is used to treat fever and that of leaves and roots is used to check inflammation, infections and to treat jaundice. Durian thrives in humid, equatorial climate with short or no dry season. An annual rainfall of 200 cm is a minimum requirement, but heavy rains prior to flower initiation affect normal flowering, and the production decreases. If minimum temperature falls below 8oC, the tree suffers from cold injury. Usually the growers propagate durian by seeds and their off springs vary from generation to generation due to genetic heterozygosity. However, the recommendation would be to go for vegetative propagation of proven elite trees. Grafted durian trees are precocious and produce crop in 4-5 years, while seedlings require as long as 10 years. The fruiting is clearly seasonal, available 2 times in a year. Immature fruits are picked only for use as 'vegetables'. Fruits take about three and a half months to reach maturity (Nath et al., 2009). When mature, the fruits drop but it should be harvested since the fallen fruits do not keep long. Harvesting is done manually or with the help of bamboo poles when the fruits are smooth, flat with far apart spines. The yield varies with the age, variety, and agro-climate of the region. However, 100-120 fruits/tree are considered as good.

Constraints for Commercial Cultivation of Exotic Fruits in India

- Lack of standard varieties suitable for the region and propagation techniques.
- Unavailability of quality and healthy planting materials.
- Technology for crop husbandry and canopy management is lacking.
- Lack of technological support for value addition through processing.
- Poor recognition of some of these crops in horticultural promotion programmes.

Strategies for Ensuring Exotic Fruits as Mega Sector

To establish exotic fruits as mega sector in Indian agricultural system, domestication and standardization of cultivation practices, facilitation for supply of planting material and increasing the demands for the produce by exploring their uses, creation of awareness among consumers and establishing a good distribution network are very crucial at the moment.

Some of the points have been enumerated here which can help ensure exotic fruits as mega sector in Indian fruit industry in coming years.

- Establish a database of existing information on climatic suitability within the country, stress tolerance and nutritive values of exotic fruit crops in India.
- Domestication through homestead cultivation of potential exotic species needs to be encouraged.
- Creating awareness about the importance of diets with nutritional value and possible role of exotic fruit species.
- Selection of superior genotypes/varieties for diverse uses like fruit juice, wine and extraction of natural colour etc.
- Standardization of propagation techniques for true-to-type and faster multiplication.
- Production of quality plants and promotional schemes for area expansion of some exotic fruit crops.
- Development of crop husbandry and canopy management techniques suitable for Indian condition.
- Development or adoption of standard protocol for post-harvest management and value addition.

Conclusion

The geo-climatic condition of India offers an excellent scope for growing different types of horticultural crops including fruits, vegetables, spices and plantation crops, medicinal and aromatic plants of high economic value. A wide range of tropical, sub- tropical and semi-temperate exotic fruits are being grown sporadically in different parts and regions of the country.

Many exotic fruit crops have immense potential to contribute in location specific fruit production chain, as they are well adapted to existing and adverse environmental conditions. Therefore, increased production of these crops will not only reduce the dependency of livelihood improvement and nutritional security on few crops species but also help in diversification of agriculture and satisfying multifarious human needs. These crops also has potential to provide many fold employment opportunities and promote agro based industries such as packaging, storage, preservation, canning and product diversification. The promotion of such species is need of the hour and systematic approach needs to be developed and an effective mechanisms need to be put in place for these crops at various level involving different stakeholders in this direction.

References

Jacob, J., Lukose, S and Sebastien, V.C. (2013). Rambutan. Published by Homegrown Nursery and Farms, Kerala. pp 1-7.

Nath V., Pandey V., Pandey., D. and Kumar, D. (2009). Fruits for the future. Volume 2- Lesser known tropical and subtropical fruits. SSPH, New Delhi. pp 554.

Patel, R. K., Singh, A., Deka, B. C. and Ngachan, S. V. (2008 a). Handbook of fruit production. ICAR Research Complex for NEH Region, Umiam-Meghalaya.

Patel, R.K., Singh, A., Yadav, D.S. and De, L.C. (2008 b).Underutilized fruits of North eastern region, India. *In: Underutilized and Underexploited Horticultural crops, Vol. 4* (ed. K.V. Peter), NIPA, New Delhi. pp 223-238.

Patel, R.K., Verma, V.K., Deshmukh, N.A., Jha, A.K., Nath, A. and Patel, R.S. (2013). Underutilized fruits of Meghalaya: An option for food and nutritional security. *In: Horticulture for Economic prosperity and Nutritional security in 21st Century.* (ed. Hazarika and Nautiyal), Westville Publishing House, New Delhi. pp 256-263.

Varghees, L. and Lukose, S. (2013). Pulasan. Published by Homegrown Nursery and Farms, Kerala. pp 1-2.

Chapter 6

Technologies in Arid Horticulture: Challenges and Opportunities

Prof. (Dr.) P. L. Saroj
CAR-Central Institute for Subtropical Horticulture, Rehmankhera-226101, Lucknow (UP)
Email: plsaroj@yahoo.co.in

Indian arid region is spread over 39.54 million ha including 31.71 million hectare as hot arid region mainly in the states of Rajasthan, Gujarat, Andhra Pradesh, Punjab and Haryana and 7.83 million hectare as cold arid region in the states of Jammu and Kashmir and Himachal Pradesh (Saroj and Dhandar, 2004). The major area under hot arid region of western Rajasthan consist of the districts of Jaisalmer, Barmer, Bikaner, Hanumangarh, Sri Ganganagar, Churu, Jhunjhunu, Sikar, Nagaur, Jodhpur, Pali and Jalore covering an area of nearly 1,43,842 sq.km. The arid zone is characterized by extremes of temperature, low and variable precipitation and very poor in soil fertility, which limit the scope for good productivity of agriculture in general and horticulture in particular. Besides moisture regimes; thermal regime and high wind velocity during crop growing period also influence crop productivity significantly. The water resources in arid region are limited and can irrigate hardly 4 per cent of the area. The irrigation water resources in the region are seasonal rivers and rivulets, surface wells and some runoff water stored in *tanka* and *khadins* and canal irrigation in hot arid region. Apart from this, in western parts of arid region, occurrence of frost is also common features during winter season which affects plants vegetative growth, productivity and quality of produce especially in aonla, lasoda, tamarind, bael and ber. However, these conditions favour physiological adaptive horticultural crops and their varieties of date palm, ber, pomegranate, citrus and vegetables such as cucurbitaceous crops, spices and some medicinal plants.

Under hot arid conditions, existing flora including fruit crops have developed and/or modified their organs to perform certain vital physiological functions such as strong deep root system (ber, bael, aonla, wood apple, jamun, etc.), synchronize their flowering and fruit development with the season of moisture availability (ker, lasora, aonla, pilu, etc.) and other xerophytic characters i.e. leaf shedding in summer (ber), scanty foliage (ker), mucilaginous sap in plant part (ker, gonda, pilu, bael, etc.), sunken stomata and fur/ hairiness and waxy coating on the leaf surface (phalsa, ber, lasoda, fig, etc.), thorny nature (ber, ker, cactus, bail etc.), and selective or reduced absorption of cation (Na+) and anions (Cl^-, $SO4^{--}$) for their survival. Therefore, cultivation of suitable species of fruits and vegetables with proper management approaches using windbreaks and shelterbelt in the arid areas will increase the sustenance of the inhabitants and provide the alternate sources of income. Rao *et al.* (1983) reported that a reduction of evaporative demand by 8-10 per cent can be achieved in leeward side of the tree shelterbelt between distances of 2 to 5 times the height of plants. Further, Gupta and Gupta (1981) reported a reduction in soil loss of 49 per cent in sheltered areas.

In spite of these constraints, the main strength of arid region are availability of vast area, surplus family labourers, ample sun shine, opportunity for exploitation of native flora, multilayer receptor crop cultivation, development of high value products, organic farming etc. Besides, wide variations in day and night temperature are very conducive for quality production of various arid fruits. Moreover, there are various researchable issues for improving productivity and profitability in arid horticulture are genotype improvement, propagation protocol development, canopy architecture engineering, crop diversification, protected cultivation, organic cultivation, ICT based technology, geo-spatial technology, postharvest management including value addition, integrated nutrient, water, weed, diseases-insect management, use of botanicals, drone application, FPO based entrepreneurship development etc.

Horticulture Scenario

In India, horticulture has made considerable progress and horticultural crops production has surpassed over total cereal production (Fig-1). India witnessed the shift in area from food grain towards horticulture crops over more than a decades (2012-13 onwards), but consumption of underutilized fruits as such is very limited, though they are climate resilient, good source as functional food and suitable for preparation of various value added products. Still in our country, 70 % Indian population consume < 50% of recommended dietary allowances of micronutrients. About 50% children are below 5 years age are under weight and 20% of them severely malnourished. Similarly, 57 % women and children suffer due to vitamin A deficiency and often rural women are anemic. At global hunger index, out of 116 country India's position is 101 (http/www/GHI). Therefore, horticulture industry has to travel a long way both in domestic and global trade. Also focused research is required in non-traditional areas like arid region.

Arid Fruits and Vegetables

Indian arid zone blessed with large number of fruits and vegetables as given in table 1. Among fruits like date palm, pomegranate, ber, aonla, kinnow, sweet orange, guava, tamarind and custard apple are grown commercially while fruits like bael, jamun, wood apple, karounda, mulberry, phalsa, ker, lasoda, jangal jalebi etc are still underutilized. Similarly in arid vegetables, tomato, brinjal, chilli, bottle gourd, cucumber, water melon, peas, cluster bean, moth bean, radish and onion are grown on large scale while kachri, snap melon, mateera, round melon, spine gourds etc are grown at limited scale. However, high cost of inputs and lack of enough incentives for production of quality varieties produce/species, product diversification, value addition, etc. also hinder crops development in such resource poor areas. The situation becomes further difficult in view of the large number of small holdings devoted to these crops which are essentially owned by weaker section, who have limited resource to invest, also cannot afford burden of credit, even if available. Marketing of arid fruits and vegetables is a major constraint in the production and disposal system and has a major role to play in making the industry viable.

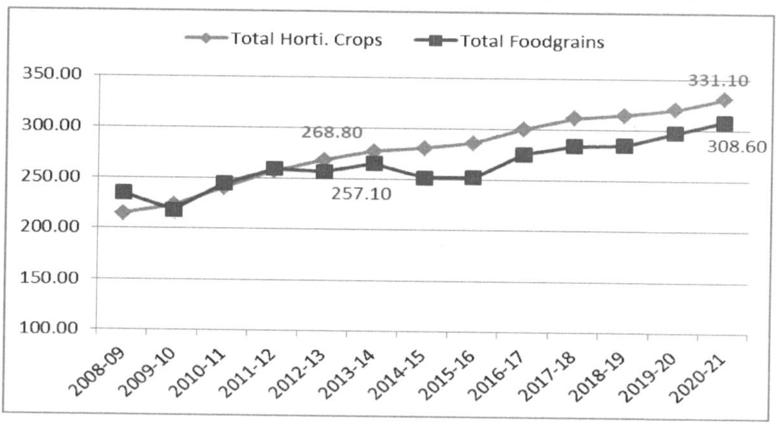

Fig. 1: Production of Horticulture *vis-a-vis* Food grains

Table 1: Fruits and vegetables of arid region

Common Name	Scientific Name	Family	Chromosome number (2n)	Origin
Arid fruits				
Aonla	*Emblica officinalis*	Euphorbaceae	18, 28	Peninsular India
Ber	*Ziziphus mauritiana*	Rhamnaceae	24	Indian Sub-continent
Bael	*Aegle marmelos* L.	Rutaceae	36	Northern India
Phalsa	*Grewia subinaequalis*	Tiliaceae	36	India
Wood apple	*Feronia limonia*	Rutaceae	18	India, Sri Lanka
Jamun	*Syzygium cumini*	Myrtaceae	40	Indian Sub-continent
Karonda	*Carissa carandus* L.	Apocynaceae	12	India
Kinnow	Hybrid of 'King' (*Citrus nobilis*) and 'Willow Leaf' (*Citrus deliciosa*)	Rutacae	18	California

Common Name	Scientific Name	Family	Chromosome number (2n)	Origin
Sweet orange	*Citrus sinensis*	Rutaceae	18	South East Asia
Guava	*Psidium guajava* L.	Myrtaceae	22, 33	Brazil
Custard apple	*Annona Squamosa* L.	Annonaceae	14	Tropical America - Peru
Date palm	*Phoenix dactylifera*	Palmaeae	36	Persian Gulf, Iraq
Pomegranate	*Punica granatum* L.	Punicaceae	16, 18	Iran
Tamarind	*Tamarindus indica*	Leguminosae	24	Tropical Africa
Lasoda	*Cordia myxa*	Boraginaceae	48	India
Mulberry	*Morus alba* L.	Moraceae	28	Northern China
Fig	*Ficus carica* L.	Moraceae	26	South East-Asia
Chironji	*Buchanania lanzan*	Anacardiaceae	22	Western Peninsula
Ker	*Capparis decidua*	Capparidaceae	14, 16	Arid region of India
Pilu	*Salvadora oleoides*	Salvadoraceae	24	India/Tropical Africa
Arid vegetables				
Tomato	*Solanum lycopersicum* L.	Solanaceae	24	South America
Brinjal	*Solanum melangena*	Solanaceae	24	Southern India
Chilli	*Capsicum annuum* L.	Solanaceae	24, 26	Central and South America
Bottle gourd	*Lagenaria siceraria*	Cucurbitaceae	22	Sub-Saharan Africa
Cucumber	*Cucumis sativus* L.	Cucurbitaceae	14	India
Water melon	*Citrullus lanatus*	Cucurbitaceae	22	Africa and India
Peas	*Pisum sativum*	Leguminosae	14	Southern Europe & Western Asia
Onion	*Allium cepa* L.	Alliaceae	16	South East Asia
Khejri	*Prosopis cineraria*	Leguminosae	28	Thar desert of India
Kachri	*Cucumis callosus*	Cucurbitaceae	14	Thar desert of India
Snapmelon	*Cucumis melo* var. *momordica*	Cucurbitaceae	24	Tropical regions of Africa and Asia
Mateera	*Citrullus lanatus*	Cucurbitaceae	22	Indian Thar desert
Bitter apple	*Citrullus colocynthis*	Cucurbitaceae	22	Indian Thar desert
Arya	*Cucumis melo* var. *chate*	Cucurbitaceae	24	Indian Thar desert
Round melon	*Citrullus vulgaris* var. *fistulolus*	Cucurbitaceae	24	India
Spine gourd	*Momordica dioica*	Cucurbitaceae	28	India
Jhaar karela	*Momordica balsamina*	Cucurbitaceae	22	Africa
Cluster bean	*Cyamopsis tetragonoloba*	Fabaceae	14	India

Common Name	Scientific Name	Family	Chromosome number (2n)	Origin
Moth bean	*Vigna aconitifolia*	Fabaceae	22	India, Pakistan and Burma
Phog	*Calligonum polygonoides*	Polygonaceae	54	Indian Thar desert
Khimp	*Leptadenia pyrotechnica*	Apocynaceae	22	Africa, Indian Thar desert

Arid fruits are very rich in nutritional and medicinal properties. Underexploited arid fruit crops such as bael, phalsa, wood apple, fig, mulberry, lasoda, kair and in arid vegetables such as kacri, mateera, bitter apple, arya, round melon, spine gourd, jhar karela, cluster bean, moth bean, phog, khimp, ivy gourd, Indian aloe and khejri are rich sources of energy and nutrients particularly micronutrients and vitamins. The micronutrients present are minerals (like iron and calcium) and vitamins (like vitamin B complex, C, folic acid, and carotenoids) while, the macronutrients present are complex carbohydrates/fiber. They contain plentiful amounts of iron, calcium, vitamin C, folic acid, carotenoids (vitamin A precursors) and phytochemicals. In Yunani and Ayurvedic formulations (eg. Triphala, chyawanprash and joshanda), they are being used since centuries for treatment of diseases and disorder by remedial methods. Some therapeutic uses of arid fruits are given in table-2.

Table 2: Some therapeutic uses of arid fruits

Common Name	Botanical Name	Therapeutic use
Aonla	*Phyllanthus officinalis*	Anti-anemic, boost immunity, improves liver functioning, healthier skin, minimise cardiovascular problem.
Bael	*Aegle marmalos* Corr.	Appetizer, stomachic, cooling, restore vitality.
Jamun	*Syzygium cuminii* L.	Stomachic, Anti-anemic, improves hemoglobin in blood, excellent fruit for diabetic patient.
Pomegranate	*Punica granatum* L	Super food ageist melignated cells, improve hemoglobin, anti-anemic, treatment of cardiovascular disease, osteoarthritis, rheumatoid arthritis.
Phalsa	*Grewia subinaequalis* D.C.	Blood purification, anemia, beneficial in blood and liver disorders.
Date palm	*Phoenix dactylefera* L.	Antibacterial, antifungal and anti-proliferative properties, good for teeth due to florin rich.
Ber	*Ziziphus nummularia*	Helps in blood purification and improves digestion.

Common Name	Botanical Name	Therapeutic use
Ker	*Capparis decidua Edgew*	Biliousness, asthama, inflammations, fever, laxative, cough, cure stomach pain, vomiting, arthritis, diabetes and hypertension.
Kaith	*Feronia limonia* L.	Stomachic, stimulant, carminative, tonic for digestive disorders and antiscorbutic.
Fig	*Ficus carica* L	Diuretic, laxative, expectorant and antispasmodic treatment of leucoderma.
Tamarind	*Tamarindus indica*	Useful in heart care and against stone formation in urinary system.
Lasoda	*Cordia dichotoma*	Anticlimactic, diuretic, expectorant.
Karonda	*Carissa carandas*	Anti-scorbutic, anemia .
Pilu	*Salvadora persica*	Rbeumatism pains, cure piles, cough and purgative, root bark is vesicant, fruit is useful in enlarge spleen and low fever with skin diseases, etc.

(Source: Goyal and Sharma, 2009; Pareek et al., 1999; Gopalan et al., 2004; CSIR, 2003)

Research and Development in Arid Horticulture

Genetic resource management: Systematic exploration programme was taken up at ICAR-CIAH, Bikaner and also centers of AICRP on Arid Zone Fruits. So far, 1294 accessions of different arid fruits and 2899 accessions of arid vegetables have been collected and conserved at the institute while 779 accessions of fruits and 448 accessions of semiarid vegetables have been collected and conserved at Regional Station, Vejalpur (Gujarat). Besides, 3289 accessions of different arid fruits have been collected and maintained in AICRP-AZF centers. ICAR-CIAH, Bikaner is recognized as National Field Repository for Date palm, Ber, Bael, Aonla, Pomegranate, Jamun and Tamarind and also developed DUS guidelines for these crops. The collected germplasm were also characterized for various parameters and duplicates were identified. For example, in ber (*Ziziphus mauritiana*), 318 collections were made but after characterization at molecular level, the core collections was down sized to 154. Recently IC No. of 75 germplasm were obtained from NBPGR and 04 trait specific germplasm were registered (Water melon for stable andromonoecious sex form and rich in carotenoid content, Monoecious line of musk melon and Tingid bug resistant in Lasoda (*Cordia myxa*). Also, determined molecular diversity of 96 accessions of pomegranate, 53 accessions of ber, 19 cultivars of citrus, 10 accessions of Jamun and 24 accessions of cow pea using 22 SSR markers. Completed DNA finger printing of 15 released varieties of arid fruits and vegetables. Identified trait specific germplasm of Date palm (Rain tolerant and high pollen production type), Bael (suitable for processing and core centric seed cavity), Pomegranate (purple fruit), Custard apple (self-fruitful and big size fruit) and Dolichos bean (high productivity and rich in antioxidant).

Development of varieties: While selection of fruit crops for arid region, one of the basic requirements is that those crops who complete their vegetative growth and reproductive phase during the period of maximum moisture availability. The fruit such as ber, pomegranate, custard apple, aonla and sour lime, conform to this prerequisite. In these crops also, there is need to select the most drought hardy cultivars for rainfed production. Early ripening cultivars seem to escape stress conditions due to receding soil moisture stored in the soil profile during the monsoon. For instances, ber cultivars Gola, Seb and Mundia for extremely dry areas, Banarasi Kadaka, Kaithli, Umran and Maharwali for dry regions, and Sanaur-2, Umran and Mehrun for comparatively humid regions have been recommended (Chadha, 2019). Apart from morphological parameters, plants should also have physiological parameters for endurance to drought for commercial cultivation in this region. Some physiological parameters identified in ber are no-mid-day depression in photosynthetic rate, low rate of transpiration, maintenance of leaf water balance, dry matter allocation, high water use efficiency, etc. It has been demonstrated that plant having capacity for drought endurance are able to maintain turgor, dry matter allocation, leaf and fruit growth even under low soil moisture level. Therefore, crop improvement programme in arid horticulture was focused on to develop high-yielding varieties and hybrids that are resistant/ tolerant to various biotic and abiotic stresses. The important fruit varieties recommended for arid region are presented in table 3.

Table 3: Recommended fruit crops and their cultivars

Fruit Crops	Cultivar
Ber	Gola, Seb, Umran, Kaithali, Banarsi Karaka, Mundia, Goma Kirti, Thar Sevika
Bael	NB-5, NB-9, Goma Yashi, Thar Divya, Thar Neelkanth, Thar Srishthi, Pant Aparna, Pant Sujata, Pant Urvashi, Pant Shivani, CISHB-1, CISHB-2
Aonla	NA-6, NA-7, Chakkaiya, Krishna, Kanchan, Balwant, Goma Aishwarya, NA-10
Date palm	Halawy, Barhee, Medzool, Khuneizi, Khalas
Pomegranate	Bhagawa, Super Bhagwa, Solapur Lal, Mridula, Jalore Seedless, Goma Khatta (for anardana)
Fig	Poona Fig, Conadria, Diana, Brown Turkey, Excel, Black Ischia, Phule Purandar
Custard apple	Balanagar, Mammoth, Island Gem, Arka Sahan, Phule Janki
Tamarind	PKM -1, Pratisthan, Yogeshwari, Goma Prateek, Anant Rudhira, Thettu Amalika
Sweet orange	Malta Blood Red, Mosambi, Satgudi, Pusa Round
Lime	Kagzi, Vikaram, Sai Sharbati, Pusa Abhinav, Pusa Udit
Karonda	Thar Kamal, Pant Sudharshan, Pant Manohar, Pant Suvarna

The ICAR- Central Institute for Arid Horticulture, Bikaner (Rajasthan) has developed 48 varieties of different arid horticultural crops, out of which 24 are in fruit crops (CIAH, 2016) which are given in Table-4. Apart from this AICRP on AZF have developed varieties of Pomegranate (Bhagwa and Super Bhagwa), Custard

Apple (Arka Sahan, APK (Ca-1) and Phule Janaki, Ber (Narendra Ber selection 1 and 2), Aonla (ND/AH-16, NA-17), Lasora (JL-7 as Karan Lasoda), Jamun (CISH J-1 & 2), *Bael* (CISH B-1 & B-2) for higher yield and quality production (AICRP-AZF at a glance).

Table 4: Varieties of fruit crops developed at ICAR-CIAH, Bikaner

Crop	Variety	Salient characteristics
Aonla	Goma Aishwariya	It is an early and drought tolerant variety
Ber	Thar Sevika	It is an early maturing hybrid (Seb x Katha) variety. Fruits are soft and juicy.
Ber	Thar Bhubharaj	It is very juicy, sweet and medium maturity variety
Ber	Thar Malti	Mid maturing group variety with high yield and soft pulp texture, no throat itching at raw stage eating
Ber	Goma Kirti	Late maturing variety but comes 10-15 days earlier than Umran and higher yielder
Bael	Goma Yashi	It is semi dwarf stature, low thorn, precocious and good yielder with smooth pericarp and suitable for commercial orcharding under dry land areas.
Bael	Thar Divya	Drought tolerant, early in maturity (second fortnight of February), precocious bearer and highly suitable for growing under dryland /rainfed hot semi-arid ecosystem.
Bael	Thar Neelkanth	Drought tolerant, lustrous growth and higher fruit yield having attractive fruit appearance under aberrant agro-climatic conditions.
Bael	Thar Shrishthi	Medium stature of the tree, locule centric variety contain high pulp and easy in processing
Bael	Thar Prakriti	Medium stature, rich in TSS and antioxidants
Bael	Thar Shivangi	Compact canopy and dwarf stature suitable for HDP and low incidence of scald
Jamun	Goma Priyanka	Medium vigorous growth habit, precocious, response to pruning and good yielder in rainfed conditions
Jamun	Thar Kranti	Early ripening (last week of May) and drought tolerant
Pomegranate	Goma Khatta	Anardana type and seed hardness is medium
Lasoda	Thar Bold	A prolific and early bearing lasoda variety suitable for preparation of vegetable, pickle and table purpose.
Chironji	Thar Priya	It is comparatively dwarf, precocious bearer and suitable for high density planting.
Khirni	Thar Rituraj	It is semi-dwarf, precocious bearer, dwarf and suitable for high density planting.
Karonda	Thar Kamal	It is comparatively dwarf, precocious bearer and suitable for high density planting and suitable for candy and jelly making.

Mahua	Thar Madhu	It is comparatively dwarf, precocious bearer and suitable for high density planting.
Mulberry	Thar Lohit	Low (frost) and high temperature tolerant colorant variety
Mulberry	Thar Harit	It is mid-season, precocious bearer, high yielding variety
Phalsa	Thar Pragti	It is dwarf, early precocious bearer (bearing in 3rd year) and drought tolerant and suitable for high density planting.
Tamarind	Goma Prateek	Long pod high yielding variety, suitable for rainfed conditions.
Wood apple	Thar Gaurav	Compact canopy, fruits rich in pectin and protein content and early maturing (1st week of November).

Similarly, ICAR-CIAH, Bikaner has developed large number of varieties of arid vegetables which are gaining popularity on farmer's field (Table 5).

Table 5: Varieties of arid vegetables developed at ICAR-CIAH, Bikaner

Crops	Varieties	Crop	Varieties
Kachri	AHK-119, AHK-200	Ivy gourd	Thar Sundari
Mateera	Thar Manak, AHW-19, AHW-65	Spinach	Thar Hariparna
Snap melon	AHS-10, AHS-82	Pompkin	Thar Kavi
Kakdi	AHC-2, AHC-13	Indian bean	Thar Kartiki, Thar Maghi
Khejri	Thar Shobha	Cluster bean	Thar Bhadavi
Bottle gourd	Thar Samridhi	Sword bean	Thar Mahi
Brinjal	Thar Rachit	Long melon	Thar Sheetal
Sponge gourd	Thar Tapish	Ridge gourd	Thar Karni
Tomato	Thar Anant	Moringa	Thar Harsha

Some basic researches carried out which have long run implications in arid horticulture are as under;

- DNA finger printing of released varieties of arid fruits like Lasoda (Thar Bold), four ber varieties (Thar Sevika, Thar Bhubharaj, Thar Malti and Goma Kirti), pomegranate (Goma Khatta); and vegetables like long melon (Thar Sheetal), ridge gourd (Thar Karni) and water melong line (YF-5-2-7) has been carried out.
- Elucidated the bottom-up mechanism of concurrent abiotic stresse tolerance/susceptibility in tolerant (Khejri, salvadora and *Z. nummularia*) and susceptible (*Z. mauritiana* and aonla) arid horticultural crops. The antioxidant defense system of above said crops was studies from January to December months (2018 and 2019) and found that the antioxidant defense system (biomolecule as well as enzymatic) of tolerant crops responded quickly to concurrent stresses and made itself competant to chelate the

resultant ROSs quickly while antioxidant defense system of susceptible crops could not responded to optimal level to chelate the resultant ROSs.

- Elucidated the bioactive compounds/phytochemicals production potential and quantified 15 major phenolic compounds in different plant parts of Phog (*Calligonum polygonoides*) and it contains 13-35 % (DW) of bioactive compounds and detected gallic acid, catechin, vanillic acid, chlorogenic acid, epicatechin, coumeric acid, catechol & syringic acid as major phenolic compounds. GC-MS/MS based phytochemicals screening of phog flower bud was carried and detected most of the phytochemicals with proved important bio-activities like anti-microbial, anti-inflammatory, anticancer, anti-diabetic, hepatoprotective, cardiovascular, antioxidant, anti-mutagenic etc. Therefore, has a great potential for molecular farming under hot arid region.
- Elucidated the seasonal effect on bioactive compounds production potential of phog foliages and found that under temperature stress (sever heat and cold) induced the bioactive compounds accumulation by more than two folds.
- Determined the nutritional and nutraceutical potential of ker, bael, ber, aonla, jamun phalsa, mulberry, drumstick and summer leafy vegetable crops along with 32 advanced guava lines developed at CHES, Vejalpur.
- Determined the molecular diversity among 24 accessions of cow pea, 96 accessions of pomegranate, 53 accessions of ber, 19 cultivars of citrus and 10 cultivars of Jamun.
- Identified one gene 'Dehydin-1' from EST database of *Z. nummularia* for drought tolerance.
- Genome-wide survey was conducted in watermelon and characterized 146AP2/ERF genes and 6 RBOH genes.
- Standardized process for somatic embryogenesis and *in-vitro* development of plantlets of two date palm cultivars viz. Halawy and Khalash.
- Studied the leaf anatomical characters in different ber genotypes (Tikadi, Mundia, Gola and Seb) in correlation with plant vigour.
- Indentified phenological growth stages of Ber, mulberry and wood apple based on extended BBCH-scale for use of optimal crop management practices.

Orchard establishment: Establishment of orchard by quality planting material of desired variety is very crucial in perennial horticulture. The planting system adopted in arid region depends largely upon land topography, soil type, fruit type and its variety. Planting is generally done in plain areas it should be done on controls in square or rectangular system, but in sloppy land terraces made across the slop. For runoff water harvesting inward micro catchment should be made. Pit bottom and side placed a clay or pond silt barrier 2-3 cm thick for reducing percolation loss of the water in light textured soil. Polythene mulch cover help in substantial reduction of evaporation loss of water and young plants should be covered for protection from high radiation and wind action. Double walled pitcher of 3-5 litre or more capacity with bottom painted use alongside of each tree with periodically filled with water helping in establishing orchard in arid region (Pareek, 1987 and Gupta *et al.*, 1987). Proper care of plant during and after planting is required. Staking of new plantations and protection from frost is also essential.

Nursery management: To boost the horticulture production, quality planting material availability play a major role. Therefore, a chain of certified nurseries are required for mass multiplication of quality planting materials. Standardization of propagation techniques for commercial orchard establishment of desired variety requires standard protocol. Establishment of mother block of newly released varieties, raising of rootstocks, use of proper growing media and polycontainers are also important considerations. Besides, use of hardy rootstocks for establishment of orchard in arid region is also important (Saroj, 2004). ICAR-Central Institute for Arid Horticulture, Bikaner has been working since 1993 and producing and distributing quality planting material to the stakeholders under different R & D activities (Sarolia *et al.*, 2018). Details of protocol standardized for propagation of the fruit crops are given in table 6.

Table 6: Methods of propagation in arid fruit crops

Fruit	Propagation method	Reference
Aonla	Patch budding in polycontainer	Saroj *et al.* 2000
Bael	*In-situ* Patch budding	Singh *et al.* 2014
Ber	*In-situ* patch budding	Nath *et al.*, 2000
Custard apple	Soft wood grafting, budding	Anon., 2011
Date palm	Off shoot (suckers), Tissue culture	
Fig	Hard wood cutting	
Mulberry	Stem cutting/ shield budding	
Phalsa	Seed and cutting	
Jamun	Shield and patch budding	Singh and Singh, 2006
Ker	Suckers/ seed	Anon., 2013
Khejri	*In-situ* Patch budding	Purohit *et al.*, 2003
Karonda	Seeds/ hard wood cutting	Bhardawaj and Sarolia, 2011
Lasoda	Patch budding	Singh et al., 2003 and Kumar *et al.*, 2022
Pomegranate	Semi-hard wood cutting under mist	Saroj *et al.*, 2008
Tamarind	*In-situ* soft wood grafting	Awasthi et al., 2005
Wood apple	*In-situ* budding	Singh and Singh, 2015

Water management: Irrigation water is a crucial input in arid region, hence judicious use of water is very essential. Scientific water management not only increases productivity of the orchard but also reduces input cost by improving water use efficiency. Besides water harvesting, mulching and weed management; water use efficiencies may be enhanced with the use of pressurized system of irrigation. Sharma *et al.* (1982-86) have reported that microcatchment slopes greater than 5 per cent did not significantly affect run off at Jodhpur and that yielded maximum when 0.5 per cent and 5 per cent slopes had 8.5 m and 7 m length and 72 m^2 and 54 m^2 catchment area per tree, respectively. Mulching with organic material have multiple benefits in reduction of moisture loss, suppress weed growth, prevent erosion and add organic matter to the soil (Gupta, 1995). In arid region black polythene mulch is very effective in ber orchard (Anonymous, 1989). Use of antitranspirant 4-6 % kaolin, 0.5-1.0 % liquid paraffin gave better result (Pareek ans Sharma, 1991). Shelterbelt and wind breaks can reduce evapotranspiration by reducing the wind velocity and stabilize micro-climate (Muthana *et al.*, 1984). Achievements in water management aspects under AICRP-AZF is in-*situ* planting of aonla in sunken basin of 5% slope plus with black polythene mulch (600 gauges) registered the highest soil moisture, morphological characters, fruit yield, chlorophyll stability index and relative water content (Aruppukottai and SK Nagar centres). In case of fig, 40% wetting zone with 900:250:275 g NPK/ tree applied through drip irrigation gave significantly higher fruit yield (Rahuri) and maximum marketable fruit yield of pomegranate was recorded with 20% wetting zone and 625:250:250 g NPK/tree in drip fertigation (Rahuri).

Integrated nutrient management: The sandy soil of the arid region is very poor in organic carbon and nitrogen content. Since, the infiltration rate of water in sandy soil is very high and water holding capacity is very poor, hence decomposition of biomass is very limited. However, regular application of organic matter not only increases its content but also improves the soil condition and makes available the various nutrients to the plant. Recommended doses of fertilizers and manures should be applied at appropriate time and according to age of the plants. Split dose of fertilizers are recommended in sandy soil of arid region. Nitrogenous fertilizers should be applied in 2-3 split doses at critical stages. Foliar feeding of nutrients such as nitrogen (0.5 to 2 % urea), zinc (0.05 to 1.0 %) and boron (0.01 to 0.05%) have given beneficial results in fruit of arid region (Pareek and Sharma, 1991). The recommended dose of manures and fertilizers for different arid fruit crops is mentioned in table 7.

Table 7: Arid fruit crops and recommended dose of manure and fertilizers

Crop	FYM (kg)	NPK $g^{-1}plant^{-1}$			Reference
		N	P	K	
Ber	50	500	250	250	AICRP-AZF, Hisar
Aonla	150	500	250	125	AICRP-AZF, Bengaluru
Pomegranate	50	625	225	225	AICRP-AZF, Rahuri (Maharashtra)
Date palm	60	600	100	70	Nath and Singh, 2003
Bael	80	500	250	250	Singh & Mishra, 2000
Fig	40	900	250	250	AICRP-AZF, Rahuri (Maharashtra)
Custard apple	25	250	125	125	Anon., 1983
Phalsa	15	100	40	25	Pundir and Pathak, 1981
Jamun	50	1000	500	500	Nath *et al.*, 2008
Karonda	10	50	100	150	Pandey and Reddy, 1989
Mulberry	30	300	150	200	Srivastava, 1984
Lasoda	25	500	250	250	Singh *et al.*, 1996

Plant canopy architecture management: Plant architecture can be maintained through proper training and pruning of fruit plants. Training at initial stages of growth gives proper shape and strong frame to the trees while pruning makes a balance between vegetative and reproductive phase. Regular pruning is an essential feature in high density planting of fruit crops. For regulation of shape, flowering and fruiting judicious removal of wood is required. The system of canopy architecture management varies with plant types, variety, age, plant vigour, planting system and growing conditions. The bushy pomegranate should be trained keeping 3-5 stems from the ground level while in other fruits (ber, aonla, jamun, citrus, bael, lasoda etc); single stem training keeping 3-4 main branches in all directions is adopted. The crops like ber and bael can be successfully trained on Y shape and espalier system for quality crop harvest and promote HDP system in these fruits (Anon., 2019 & 2020). In different fruit crops, pruning time and level are given in table 8.

Table 8: Pruning time and level under different arid fruit crops

Fruit crop	Pruning time	Pruning level	New Approaches	Reference
Ber	End of May under rainfed and in April under irrigated condition	The main shoots of the previous season are cut back retaining 15-25 nodes	Under different training systems (T, Y, Espalier) pruned by hedging and topping techniques in established plant	Saroj *et al.*, 2019
Phalsa	January in North India	Hard pruned at 20-30 cm height from ground	Trained as shrub require light pruning (25-50 % of previous year growth)	Singh and Sharma, 1961
Lasoda	December-January	Light pruning to maintain shape	Defoliation by chemicals	Chundawat, 1990
Mulberry	December-January	Medium level and for silk worm rearing hard twice a year require	Bottom pruning (0-15 cm height) for maximum leaf harvest	Nath *et al.*, 2008
Fig	Just after harvest the crop	Light to hard (2-4 bud leave previous season growth level)	Old (October flowering) and new (Spring flowering) growth	Pareek and Nath, 2006
Custard apple	Light pruning spring	25 % top branches in March-April	Bear new and old branches	Nath *et al.*, 2008

Fruit based cropping systems: Monoculture in arid zone is highly risk prone due to crop failures, hence a suitable tree crop combinations is essential for alleviating the risk, improvement in productivity per unit area/volume as a result of efficient use of natural resources and inputs, generation of better income and ameliorate and to improve edaphoclimatic conditions. Multi-storey combinations incorporating large trees, small trees and ground crops can be used based on edaphoclimatic and rainfall pattern eg., in low rainfall (300-500 mm) zone- combinations such as *khejri* or ber + ber or drumstick + vegetables (legumes and cucurbits); in 500-700 mm rainfall zone-combination of mango or ber or aonla or guava + pomegranate or sour lime or lemon or drumstick + solanaceous or leguminous or cucurbitaceous vegetables; and in 700-1000 mm rainfall zone-combination of mahua or palmyrah palm or tamarind or guava + sour lime or pomegranate or aonla + vegetables can be adopted (Pareek and Awasthi, 2008). In multi crop combinations, maximum benefit: cost ratio of 3.48:1 was obtained when crops were grown in combination of aonla + ber + karonda + moth bean + mustard. Among the perennial components, higher benefit: cost ratio (2.22:1) was found under ber. In this cropping system, the flow of cash was for almost nine months of the year under arid condition of Rajasthan (Krishna *et al.*, 2018). Under hot arid condition, ber cultivar Gola as over storey component and cluster bean, mustard and Indian aloe can be integrated into the system as compatible ground storey component as compared to groundnut wheat which is dominated rotation under irrigated hot arid condition (Saroj *et al.*, 2003).

In arid climate, highest net profit was obtained in aonla based cropping system intercropped with moth bean-cumin system followed by moth bean-chickpea cropping system (Saroj and Krishna, 2017). The tree canopy of aonla allows filtered light and permit intercropping even after it has made full growth. Among the two models developed at ICAR-CIAH, Bikaner; highest gross return was obtained under aonla–*Prosopis*– sehjana –moth bean–mustard model (Awasthi *et al.*, 2005). *Khejri* (*Prosopis cineraria*) is an important component of farming system in hot arid region. This tree grows well in all sorts of climatic constraints which is evidenced by the fact that new foliar growth, flowering and fruiting occur during severe dry months (March-June) when most other trees of the desert remain leafless or in dormant condition. In *khejri* based cropping system ber, *lasoda, pilu* and *ker* can be grown as filler crops (Kaushik and Kumar, 2003). Agri-horticultural combinations with legume intercrops and horti-pastoral system is beneficial with the system (Subbulakshmi *et al.*, 2016) could have combinations such as *khejri (Prosopis cineraria)*+ber+dhaman *(Cenchrus ciliaris, C. setigerus) or* sewan *(Laisurus sindicus).*

Integrated pest management: In arid region, insect-diseases complex is less prominent but due to increase in vegetation and cropping intensity, some major biotic stresses are causing significant loss in horticultural crops of this region. Major pest and diseases in arid horticultural crops are as under.

Insect-pest	Termites, fruit fly (ber), fruit borer (ber), pomegranate butterfly, scale insect (date palm), stone weevil (ber), blue butterfly (ber), bark eating caterpillar, lemon butterfly (citrus), gall midge mites pomegranate, date palm), etc
Diseases	Powdery mildew, (ber), wilt (citrus and bael), fungal leaf spot (bael, date palm, ber, etc.), bacterial leaf spot & fruit spot (pomegranate and bael), anthracnose (aonla and pomegranate), blight (pomegranate), mosaic and leaf curl viruses.

Therefore, an integrated pest and disease management strategies to be required to control the insect–pest and diseases below economic threshold level (ETL). Field sanitation, deep summer ploughing, hoeing-weeding, destroying infested fruits *etc* and other crop specific good cultural activities will definitely reduce the major insect-pest and disease load from the field. Growing non-host crop and removal of infected crop debris, periodical removal of webs around affected portion at the time of pruning (ber, aonla, anar) against fruit borer and bark eating caterpillar, bagging of fruits (pomegranate) for fruit fly and cue-lure traps (date palm) with growing available resistant genotypes are other important approaches. Various label claim formulations Dimethoate (0.1%), Thiamethaxam-25%WG (0.05%), Indoxicarb (0.07%) or Imidacloprid (3-5 ml/10 lit. of water) used for controlling different pests. Apart from insecticides under organic module can be used as bio-pesticides namely, neem seed kernal extract (NSKE @ 2.5-5%), neem oil, tumba fruit extract (TFE 5% or Thar Jevik) on various crops effective in controlling aphid, thrips, jassids, butter flies, fruit flies, moths, *etc.* in pomegranate, citrus and aonla.

Similarly, different fungicides can be used for controlling different diseases like; powdery mildew can be controlled by wettable sulphur (0.25%) and kerathane (0.1%). Anthracnose by Spraying of Carbendazim (0.1%) or Difolatan (0.2%), Leaf spot by two fungicidal sprays of mancozeb or copper oxychloride @ 0.2% at 15 days interval, bacterial leaf spot by spraying of 200 ppm streptocycline can be controlled followed by Pausamycin (0.05%) + Copper oxychloride (0.2%) with 3 sprays at fortnightly intervals. In organic way through biocontrol agent based on *Trichoderma, Aspergillus, Pseudomonads, etc.* isolates and some herbal extracts of *Aloe vera*, turmeric, black pepper, neem kernel extract can be used effectively against the pathogens (Anon., 2019).

Post-harvest Management

The horticultural produce suffers heavy post-harvest losses in the absence of adequate post-harvest and marketing infrastructure such as pre-cooling units, packing and grading sheds, short and long term cold storage facilities, refrigerated containers, storage and phyto-sanitary facilities etc. The post harvest losses are very high (20 to 40%) at different stages of grading, packing, storage, transport and finally marketing and reaching to consumers doorstep. Therefore, post harvest handling and value addition of perishable commodities is needed to achieve better price of produce in the market. Arid fruits can be used for making various value added products (Table-9) as drying, brining, pickling, beverage making, preserve making, *etc.* The date palm harvested at doka stage blanched 10-25 minutes was required for making dry dates (*chhuhara*). For making soft dates, curing for much shorter period i.e. for 20-25 seconds followed by dehydration at 30-40^{0}C was required (Chandra *et al.*, 1994). Dipping of ker fruits in 4-6 per cent salt solution or 0.2 per cent NaOH solution for about a week improves the quality of dehydrated product. Date biscuits prepared by adding 10 and 20 per cent date pulp powder were found nutritive in terms of protein and sugars (Singh *et al*, 2013). Proper packaging is an essential prerequisite to reduce losses of produce and attract the consumers.

In arid region, intense solar radiation is available during major part of the year; therefore, sun drying is very common and easy method. For sun drying in open, produce is spread in single layer on *pucca* floor or bamboo structure, or wooden crates or metal trays depending upon availability, convenience in handling, kind of fruits and the product type. Various fruits like ber, date palm, aonla, fig, pomegranate and *ker* etc. are being dried in sun. In order to maintain better hygienic conditions and quick drying, recently mechanical dehydration like solar, osmotic, freeze, fluidized bed drying, *etc.* have been developed. Pre-drying treatments like blanching or brining or sulphitation improves the quality of dehydrated products.

The air circulating type electric dehydrators have precise temperature control system inside the chamber, which yields desired quality product without spoilage.

Table 9: Value addition of arid fruit crop

Crops	Value added products
Fig	Dehydrated dry slices, confectionary products, nuggets, jam
Tamarind	Jam, jelly, syrup, pulp powder, candy, seed kernel powder
Wood apple	Chutney, pickle, frozen puree, sauce
Custard apple	Puree, jam, RTS, juice, frozen pulp
Jamun	Jam, juice, RTS, vinegar, preserve, frozen pulp
Ber	Candy, syrup, churan, bar, toffee
Pilu	Beverage drinks, RTS, squash
Date palm	Pind, chhuhara, biscuit, wine, jaggery
Pomegranate	Juice, squash, RTS, RTE (aril ready to eat), anardana
Aonla	Candy, preserve, toffee, juice, supari, powder, pickle, shreds
Bael	Nectar squash, leather, slab, powder, candy

(Source- Goyal and Sharma, 2009)

Human Resource Development

Human resource development at different stages is very essential for upgrading their knowledge and skill. For quick dissemination of technologies and adoption of GAP, the capacity building progammes may be taken up at regular intervals. The extension personnel should be trained as per need of the farmers and site specific practical modules should be included as actual field conditions. The extension activities like; front line demonstrations of improved arid horticultural technologies at farmers fields, farmers trainings (on/off campus) or farmers school, farmers meetings-interaction-visits, method demonstrations, practical exposure of techniques to end users, farmers group organization, activities for empowerment of farm women, *etc.* should be done to encourage the technological intervention for promotion of arid horticulture.

Area Expansion in Arid Horticulture

In arid region, considerable area has come up under fruits like ber, pomegranate, bael, fig, aonla, custard apple, date palm, karonda, lasoda, tamarind, phalsa etc in different parts of the country. This has become possible as a result of the research and developmental efforts. There is need to develop large orchards of under-utilized fruits like jamun, mahua, chironji, khirni, wood apple, fig, karonda, lasoda, manila tamarind *etc.* as these fruits are of high nutritive value and are rich in anti-oxidants. As an estimate, the area and production under arid fruits will be 7.65 lakh ha and 96.35 lakh tones respectively up to 2030, which was only 1.42 lakh ha and 14.29 lakh tones during 1993-94 (Table 10). The Growth rate of Arid Horticulture is equal to National level. Productivity of Onion, Potato and Sweet potato is better than National productivity. Pomegranate Cv. Bhagava alone occupied 86.1% area in the country. Similarly, Aonla Cv. Neelam (NA-7) covers 74% area. Kachari Cv. AHK-119 occupied 67.87% area in the arid region. The state of Rajasthan is the largest producer of Hinna (98%), Isabgol (95%), Guar (80%) and Coriander (65%) in the country. As per Third Party Report (2020) appointed by ICAR, the area under arid horticulture has increased about 15-17% in last 8 years.

Table 10: Area and production of arid fruit crops.

Crops	1993-94		2016-17		2030	
	Area (ha)	Production (MT)	Area (ha)	Production (MT)	Area (ha)	Production (MT)
Ber	41,256	3,30,048	85,500	8,95,000	1,10,000	12,10,000
Pomegranate	4,500	45,000	2,02,050	26,68,950	3,50,000	50,00,000
Aonla	26,000	2,86,000	75,000	8,25,000	1,00,000	13,00,000
Date palm	5,000	41,000	25,000	2,00,000	40,000	3,75,000
Sapota	49,000	6,44,000	80,000	9,13,000	1,10,000	12,80,000
Tamarind	10,500	52,500	15,280	82,300	20,000	1,20,000
Others (Custard apple, Fig & Phalsa)	6,600	30,450	15,700	1,31,700	35,000	3,50,000
Total	1,42,856	14,28,998	4,98,530	57,15,950	7,65,000	96,35,000

Way Forward

- In the scenario of climate change, more focus on climate resilient genetic improvement and varietal development will definitely open the new avenues.
- Specialty commodities such as off-season varieties, cold and heat tolerant varieties, multiple disease resistant varieties using advance breeding approaches of high throughput technology is the need of hour.
- Adoption of modern nursery practices and ensuring availability of quality seed and clean planting materials.
- Development of effective 'Weather Forecasting Model' for management of frost/ hail storm as well as pest and diseases.
- Formulation of novel bio-pesticides using arid flora and exploitation of native rhizo-microbiome as eco-friendly approach.
- Utilization of abundant solar radiation for various operations and post-harvest management.
- Promotion of protected cultivation, peri-urban horticulture, container planting and integration of off-farm enterprises.
- Farm mechanization, use of sensor based technologies and drone application in arid horticulture.
- Exploiting the market potential of naturally organic produce.

◈ Skill development for diversification of horticulture in dryland areas.

References

Anonymous (1983). Biennial Report: 1981-82. AICRP on Arid Zone Fruits, MPKV Rahuri (Maharashtra): 29-40.

Anonymous (1989). Proceedings of fifth National Workshop on Arid Zone Fruit Research. Gujarat Agricultural University, Sardarkrushinagar (July 6-9), AICRP on Arid Zone Fruits, Coordinator Cell, ICAR, Tech. Doc., 27.

Anonymous (2011). Vision-2030, ICAR-CIAH. Mother plants and propagation techniques standardization : 12

Anonymous (2019 & 2020). Annual Report, ICAR-CIAH, Bikaner: pp.30-38 & 28-31

Awasth, O. P., Saroj P. L. and Dhandar, D. G. (2005). Standardization of time on success of patch budding in tamarind (*Tamarindus indica* L.) under arid conditions. *Progressive Horticulture*, 37 (2): 294-297.

Awasthi, O. P., Singh I. S. and Dhandar, D. G. (2005). Aonla based multi-tier cropping system for sustainable production in western Rajasthan. National Seminar on Globalization of aonla. Aonla Growers Association of India, Madurai, Tamil Nadu: p.44.

Bhardwaj, R. L. and Sarolia, D. K. (2011). Modern Nursery Management, Agrobios (India) (ISBN-978-81-7754- 434-3): 1-403.

Bhargava R., Maheshweri, S. K., Haldhar, S. M. and Sharma, S. K. (2014). Management strategies for adverse climate impacts and abiotic stress in arid horticultural crops. ICAR- Central Hnstitute for Arid Horticulture, Bikaner. CIAH/Tech./ Pub. No. 52: 42.

Chadha, K. L. (2019). Ber. In: Hand Book of Horticulture, II revised edition Part-1, DKMA, ICAR, New Delhi: p271.

Chandra, A., Chandra Anju and Gupta, I. C. (1994). Arid Fruit Research, Scientific Publisher, Jodhpur: pp. 250-54.

Chundawat, B. S. (1990). Arid Fruit Culture. Oxford and IBH Publishing Co. Pvt. Ltd., New Delhi: pp.134-38.

CIAH (2011). ICAR-CIAH at a glance. Bikaner, India

CSIR (2003). Council of Scientific and Industrial Research. The Wealth of India, New Delhi, India: p.140.

Goyal, M. and Sharma, S K. (2009). Traditional wisdom and value addition prospects of arid foods of desert region of North West India. *Indian J. Traditional Knowledge*. 8(4):581–585.

Gupta, I.C., Singh P.M., Yadav N.D. and Sharma B.D. (1987). Paudho ko susthapit karne ke naye vidhi. Pariwardhit gamala. *Awiskar*, 17 (8): 293-294.

Gupta, J.P. (1995). Water losses and their control in rainfed agriculture. In: Sustainable development of dryland agriculture in India (Edt. Singh R.P.). Scientific Publisher, Jodhpur: pp.169-176.

Gupta, J.P. and Gupta, G. N.(1981). A note on wind erosion from a cultivated field in western Rajasthan. *J. Indian Soc. Soil Science*, 19 (2): 278-279.

Kaushik, N. and Kumar Virendra (2003). Khejri based agroforestry system for arid Haryana. *Journal of Arid Environment*, **55**:433-450.

Krishna, H., Saroj, P. L., Jatav, M. K. and Awasthi, O. P. (2018).Fruit based cropping models to increase farmer's income-An experience of arid region. ICAR-Central institute for arid horticulture, Bikaner. CIAH/Tech./Pub. No. 70: 61.

Krishna, Hare and Singh D. (2013). Micro-propagation of lasora (*Cordia myxa* Roxb.). *Indian J. Hort.*, **70**(3): 323-327.

Kumar, K., Sarolia, D.K., Saroj, P.L., Samadia, D.K. and Singh, D. (2022). Standardization of seed, rootstock and plant standard of lasoda (*Cordia myxa* L.) for commercial orcharding in India. *Journal of Environmental Biology*, 43:267-275.

Muthana, K.D., Yadav M.S., Mertia, R.S. and Arora, G.D. (1984). Shelterbelt plantation in arid regions. *Indian Farming*, ICAR: 33-37.

More, T. A., Samadia, D. K., Awasthi, O. P. and Hiwale, S. S. (2008). Varieties and Hybrids of CIAH. Tech. Bull. No. 30, Bikaner: p.11

Nath V., Kumar Dinesh and Pandey V. (2008). Ber, Jamun, Custard apple, Fig, Pilu. In: Fruit for the future, Satish Seril Publishing House, New Delhi:pp. 60-61, 273, 294, 301-322 & 438.

Nath V. and Singh, R. S. (2003). Boost up date palm production in arid region. *Intensive Agriculture*. **41**(7-8):3-6.

Nath V., Saroj, P. L., Singh, R.S., Bhargava, R. and Pareek, O.P. (2000) *In-situ* establishment of ber orchard under hot arid ecosystem of Rajasthan. *Indian J. Hort.*, 57 (1): 21-26.

Pandey R M and Reddy Y N. (1989). Nutritional requirement for karonda. *Intensive Agriculture*. 26 :6-15.

Pareek, O. P. (1987). Contingency plan for fruit production under drought conditions. In: Fourth National Workshop on Arid Zone Fruits proceeding, Agriculture College and Research Institute, Madurai, Tamil Nadu (16-19 Sep.). Tech. Doc. No. 20.

Pareek, O. P. and Awasthi, O. P. (2008). Horticulture based farming systems for arid region. In: Diversification of arid farming system (Eds. Pratap Narayan *et al.*). Arid zone Research Association of India and Scientific Publisher, Jodhpur: pp.12-22.

Pareek, O. P. and Nath Vishal (1996). Fig. In: Co-ordinated fruit research in Indian arid zone-Two decadess profile. NRC for Arid Horticulture, Bikaner: pp1-105.

Pareek, O. P. and Sharma, S. (1991). Fruit trees for arid and semi-arid lands. *Indian Farming*, 41:25-30.

Purohit, A. K., Samadia, D. K. and Pareek, O. P. (2003). Studies on growth enhancement of Khejri (*Prossopis cineraria* L.) seedlings and *in-situ* budded plants. *Annals of Arid Zone*. 42(1): 69-74.

Rao, G.G.S.N., Ramana Rao, B.V. and Ramakrishna, Y. S. (1983). Wind regime and evaporative demand as influenced by shelterbelts. *IDOJARAS*. 87 (1):14-21.

Samadia, D. K. (2016). Horticulture based crop production site management approach (HBCPSMA). *Journal of Agriculture and Ecology*, **1**: 1-9.

Saroj, P.L., Vishal Nath and Vashishtha, B.B. (2000). Effect of polycontainers on germination, seedling vigour, root characters and budding success in aonla. *Indian J. Hort.*, 57 (4): 300-304.

Saro,j P.L. and Krishna H. (2017). Enhancing resource use efficiency through intercropping of seed spices with arid fruit crops. In: Seed spices for enhancing farmers prosperity and livelihood security (Book of lead paper & abstract). ICAR-NRC SS, Ajmer: pp 36-41.

Saroj, P. L., Dhandar, D. G., Sharma, B. D., Bhargava, R. and Purohit, C. K. (2003). Ber based agri-horti system: A sustainable land use for arid ecosystem. *Indian Journal of Agroforestry*, 5(1&2): 30-35.

Saroj, P. L., Awasthi, O. P., Bhargava, R., Singh, U. V. (2008). Standardization of pomegranate propagation by cutting under mist system in hot arid region. *Indian J. Hort.*, 65: 25–30.

Saroj, P. L. and Dhandar, D. G. (2004). Changing scenario in arid horticulture. In: Advances in Arid Horticulture, Vol.1. (Editors: P.L. Saroj *et al.*) Pub. International Book Distributing Company, Lucknow (UP). Pp.1-27.

Saroj, P.L. (2004). Nursery management in arid ecosystem, In: Advances in Arid Horticulture, Vol.1. (Editors: P.L. Saroj *et al.*) Pub. International Book Distributing Company, Lucknow (UP). Pp. 233-251.

Sarolia, D. K., Meena, R. K., Kumar Ramesh, Singh, P. P. and Gora, J. S. (2018). Mass multiplication of quality planting material in horticulture crops. In: *Horticulture in arid and semi arid region* (Eds: Jatav M.K., Saroj P.L. and Sharma B.D.). New India Publishing Agency, New Delhi: Pp. 305-314.

Sarolia, D. K, Samadia, D. K., Choudhary, B. R. and Singh, D. (2018). Production of quality seed and planting materials. CIAH/Tech./Pub. No. 75, Page No. 38 ICAR-Central Institute for Arid Horticulture, Bikaner (Rajasthan): 1-38.

Sharma, K.D., Pareek, O. P. and Singh, H. P. (1982). Effect of run off concentration in growth and yield of jujube. *Agriiculture Water Management*, 5: 73-84.

Sharma, K. D., Pareek, O. P. and Singh, H. P. (1982). Micro catchment water harvesting for raising jujube orchard in arid climate. *Trans. ASAE*, 29:112-118.

Singh, R. S., Nath V. and Tiwari, J.C. (1996). Lasoda- a promising fruit tree for arid ecosystem, *Deco Mirror*, **3** (2): 5-11.

Singh, D. B., Saroj, P. L. and Vashistha, B. B. (2003). Effect of time of budding in propagation of lasoda (*Cordia myxa* L). *Progressive Horticulture*, 35(2):230-232.

Singh, J. P. and Sharma, H. C. (1961). Effect of time and severity of pruning on growth, yield and fruit quality of phalsa. *Indian J. Horticulture.*, 18(1):20-28.

Singh Sanjay and Singh, A. K. (2006). Standardization of method and time of propagation in jamun (*Syzygium cuminii* L.) under semi arid environment of western India. *Indian Journal of Agricultural Sciences*, 76:142-45

Singh, A.K., Singh, S., Singh, R.S., Contractor, K. and Makwana, P. (2014). *In-situ* patch budding for better establishment of bael in rainfed areas. *Indian Horticulture*, 59(5): 24-25.

Singh Sanjay and Singh, A. K. (2015). Standardization of time of softwood grafting in mahua (*Bassia latifolia*) and khirni (*Manilkara hexandra*) under semi-arid environment of western India. *Indian Journal of Agricultural Sciences*, 85(2): 166-70.

Srivastava, R.P. (1984). Nutrition of fruit trees. ICAR publication, New Delhi: 276-77.

Subhulakshmi, V., Birbal, Soni M L. and Yadava N D. (2016). Agri-horti-silvi system: A successful crop production practices in arid region. In: Principle and Practices of GAPs for horticultural crops (Eds. Jatav *et al.*): pp. 369-382.

Chapter 7

Guava Breeding for Future Challenges

D.S. Mishra[1], A.K. Singh[1] and Jagdish Rane[2]
[1]Central Horticultural Experiment Station (ICAR-CIAH, Bikaner), Vejalpur (Godhra), Panchmahals-389340 (Gujarat)
[2]ICAR-Central Institute of Arid Horticulture, Beechwal, Bikaner-334006, Rajasthan

Guava (*Psidium guajava* L.) belongs to the Myrtaceae family; it thrives on a variety of soils of humid and dry climates (Rodríguez-Medina*et al.*, 2004). The largest guava-producing countries are Mexico and Brazil in Central and South America, respectively, and India and Thailand in Asia (Padilla-Ramirez*et al.*, 2003). Guava fruits are either used fresh or processed into juice or jelly, known for its high nutritional value as well as its medicinal uses (Medina and Pagano, 2003). It is a good source of health promoting substances like vitamins C,dietary fiber, β-carotene, and it is known for its ability to reduce LDL cholesterol and triglycerides (Setiawan *et al.*, 2001).Guava is often marketed as "super fruit", being rich in vitamins 'A' and 'C' with seeds that are rich in omega-3, omega-6 polyunsaturated fatty acids and especially dietary fibre (Nimisha *et al.*, 2013).It is cultivated throughout the country and considered to be one of the hardy fruits, capable of growing under adverse conditions of soil and climate (Chundawat, 1990). Major guava producing states include Madhya Pradesh, Uttar Pradesh, Bihar, Maharashtra, West Bengal, Punjab, Chhattisgarh, Karnataka, Gujarat and Andhra Pradesh. In India it is the fourth and fifth most important crop by area and production respectively. It occupies an area of 0.27 m ha with a total production of 4.05 mt (Anonymous, 2018). It is commercially important in many other countries viz., Australia, Bangladesh, China, Costa Rica, Cuba, Indonesia, Israel, Kenya, Malaysia, Myanmar, Nigeria, Pakistan, Panama, Philippines, South Africa, Sri Lanka, Sudan, Tanzania, Thailand, Venezuela and Vietnam (Mishra *et al.*, 2018).

Global Scenario of Guava Production

Guava is one of the most consumed fruits worldwide, with an increasing international trade over recent years (Aranguren *et al.*, 2010). The guava fruit is in demand in domestic as well as in international markets and it is traded in more than 60 countries. Major guava producing countries are India, Brazil, Mexico, South Africa, Jamaica, Kenya, Cuba, United States of America, Egypt, Thailand, Columbia and Pakistan (Singh, 2008).

The international trade of guava is currently limited to processed products which are exported to United States, Japan and Europe. The largest guava producing countries are Mexico and Brazil in Central and South America, respectively, and India and Thailand in Asia (Padilla *et al.*, 2003). India is world's leading guava producing country (40.40 %) followed by China, Thailand, Mexico and Indonesia (https://www.tridge.com/intelligences/guava/production).Brazil, Mexico, Thailand, Peru, Philippines are the major exporting countries of guava fruits and processed products (Varmudy, 2013).

Origin and Distribution

The guava is said to have originated in tropical America (Lucas *et al.*, 2005) and might have been domesticated over 2000 years ago (Mata and Rodriguez, 1990). However, the origin place of guava is uncertain, but it is believed to be an area extending from southern Mexico through Central America (Mortan, 1987). De Candolle (1904) opined that guava originated in Mexico;however, according to Purseglove (1968) it originated in Brazil. Guava was introduced in India in the early 17th century (Hayes, 1953). It is believed that the Spaniards introduced the guava tree of the Pacific to the Philippines and India, from where it reached to the Malay Archipelago, to Hawaii and to South Africa (Soubihe, 1951). However, scientific evidences suggest that guava is having a pre-historic anthropogenic distribution all over the Antilles (Newsom, 1993). Now, it is widely distributed throughout the tropical and sub-tropical parts of the world (Kwee and Chong, 1990). 'Araça' is a term used in Brazil for wild *Psidium* spp., including *P. cattleyanum* Sab., *P. incanescens* Martins, *P. gradifolium* Martins, *P. arboretum* Vell, *P. guineense* Sw. and *P. humile* Vell., that are native to South America (Raseira and Raseira, 1996).

Table 1: Global production scenario of guava (Pereira *et al.*, 2016)

Country	Area (000 ha)	Production (000 t)	Productivity (t/ha)	Cultivars	Propagation methods	Reference
Bangladesh	18.0	200.0	11.11	Kazi Peara (guava), BARI Peara, BAU Guava 1-8, Kazi, Swarupkathi, Mukundapuri,Allahabad	Stem herbaceous cuttings, air layering (>95%)	BBS, 2012, Rahim et al., 2011, Saha, 2015
India	255.0 z	4102.0	16.08	Allahbad Safeda, Sardar (L49), Lalit, Shweta, Arka Kiran	Stem herbaceous cuttings, chip budding and grafting (>95%)	NHB (2015-16), Mitra, 2014
Indonesia	9.4	208.0	22.12	Indonesian white, Indonesian seedless, ARP9409, JBT001	Stem herbaceous cuttings, grafting (>95%)	BPS, 2014, Nasution and Haditi, 2014
Malaysia	4.0	65.0	16.25	Red Malysian, GU 8 (Jambu Kamopochea), GU 11 (Giant guava), GU15 (Jade Seedless), Batu Arang, Subang 6 and Sitiawan	Stem herbaceous cuttings, grafting andair layering (>95%)	Pommer and Murakami, 2009, MAM, 2013
Pakistan	66.38	496.0	7.47	Sadabahar Gola (Round), Sadabahar Surahi (Pyriform), Large Pyriform, Larkana Pyriform and Safeda	Seed (>80%), cutting and air layering	MNFSR, 2014, Usman *et al.*, 2013
Thailand	7.94	168.0	21.15	Klomsali, Salithong and Kimchu	Soft wood cuttings, grafting (>95%)	Siwarungson *et al.*, 2013
Taiwan	7.12	168.0	23.59	Century, Tai-kuo-bar, Pearl and Crystal	Soft wood cuttings and wedge grafting (>98%)	Agriculture and Food Agency, 2014
Egypt	17.5	315.0	18.0	Bassateen El Sabahia, Bassateen Edfina and Allahabad Safeda	Soft wood cuttings and mound layering (>85%)	Pommer and Murakami 2009, El-Shereif ,2016

Country	Area (000 ha)	Production (000 t)	Productivity (t/ha)	Cultivars	Propagation methods	Reference
South Africa	0.5	12.5	25.0	TSG2, Fan Retief, Fredene, Dimple, Jonelle, Welheim, Frederika, Frank Marlhebe, Van Zyl, Rousseau and Du Preez	Soft wood cuttings, air layering and grafting (>80%)	DAFF, 2015, HORTGRO,2015
Cuba	7.31	47.88	6.55	Enana Roja Cubana		Perez-Rodríguez *et al.*, 2013, Valdés-Infante *et al.*, 2012, Pommer *et al.*, 2006
Mexico	20.44	302.7	14.81	Wild types named China and Media China	Seeds and root cuttings	Martínez-de-Lara *etal.*, 2004
Brazil	14.98	349.6	23.34	Paluma, Rica, Século XXI, Pedro Sato, Sassaoka, Kumagai and Ogawa	Stem herbaceous cuttings (> 95%)	Pereira and Nachtigal, 2009
Colombia	8.34	69.1	8.28	Regional Blanca, Regional Roja, Ráquira Blanca, Guavatá Victoria, Pera and Manzana	Grafting or cutting	Ocampo and Núñez, 2007, Parra-Coronado, 2014
Venezuela	3.50	55.7	15.9	Criolla Roja, San Miguel and Rio Chiquito	Seeds	Aular and Casares, 2011

Taxonomy

The *Myrtaceae* or the myrtle family comprises at least 140 genera and some 3800 to 5650 species. There are four genera of interest which produce edible fruits: *Psidium, Eugenia, Syzygium* and *Feijoa*. The genus *Psidium* contains about 150 species of which *P. guajava, P. cattleianum, P. friedrichsthalianum, P. guineense, P. littorale, P. acutangulum*, etc. are commercially important. The *Eugenia* is a small group in which important species are *E. brasiliensis, E. uniflora, E. victoriana, E. axillaris, E. aggregata, E. dysenterica, E. lutescens, E. luschnathiana*, etc. The *Syzygium* has about 500 species and most of them have originated and grow mainly in the south-east Asia. The important species are *S. jambos, S. malaccense, S. suborbiculare, S. paniculatum, S. acqueum, S. cordatum, S. cumini, S. forte, S. Samarangense*, etc. The *Feijoa* (*Acca sellowiana*) is grown mostly in Brazil, and Uruguay. The fruit is rich in protein, carbohydrates and fibre. The *E. uniflora* (pitanga) fruit pulp is a good source of carotene and ascorbic acid. The fruits of *P. guajava* are considered as one of the richest sources of antioxidant among the tropical fruits (Mitra, *et al.*, 2012). *Myrtaceae* is subdivided into two main subfamilies- the Leptospermoideae, which is distributed mostly in Asia and Africa, and the Myrtoideae, found in tropical America, Asia, Australia and the Pacific. Myrtoideae subfamily has fleshy fruit and opposite leaves and Leptospermoideae has dried fruit. Most of the genus of edible fruits such as the *Acca,Eugenia, Myrcia, Myrciaria, Myrcenthes, Myrtus, Rhodomyrtus, Psidium, Syzygium* comes under the subfamily Myrtoideae.

Guava belongs to the family *Myrtacaeae*, which has more than 80 genera and 3000 species, distributed throughout the tropical and subtropical parts of the world (Nakasone and Paull, 1998). Important characteristics of family are: the bark is shining grayish brown in colour which comes off in scales easily, the shoots are angular, stamens are numerous, ovary is inferior and fruit is berry with many seeds embedded in pulp (Ray, 2002). In guava leaves are simple and opposite, flowers are hermaphrodite, solitary or in two to three flowered cymes with four to five calyx lobes, separated or united at the base, four or five petals and numerous stamens, ovary is inferior with one to three or more cells, the fruit is berry and seeds may be few or many. The stigma is exerted above the stamens, which favours self-pollination without the help of external agency (Pommer and Murakami, 2009). Most of the commercial varieties of guava are diploids (2n=22), except the seedless types which are triploids (2n=33)in nature (Kumar and Ranade, 1952). The same chromosome number was reported by Raman *et al.* (1971) in a seedless variety of *P. guajava* suggesting that triploidy is the cause of seedlessness in guava.

Important Genera and Species

Psidiumguajava is cultivated species of guava; however, other genera of significance are *Eugenia, Syzygium* and *Feijoa* (Nakasone and Paull, 1998). The wild species of guava are of considerable importance in breeding programmes (Dinesh and Vasugi, 2010).

Psidium

Hayes (1953) reported the genus to contain about 150 species of evergreen trees and shrubs, though only a few have been studied in detail. In addition to *P.guajava*, some other related species which also produce edible fruits or they can be used in future guava breeding programmes are (Dinesh and Vasugi, 2010):

***P. Cattleianum var. cattleianum* (Sabine) [Syn. *P. littorale* (Raddi) var. *longipes* (Berg.)]:** It is a wild subtropical species closely related toguava, known as 'Strawberry guava 'due to the sweet aroma like strawberry. It is adaptable to different soil types and it is coldresistant too. It is a small tree or shrub with a smooth bark. Leaves are obovate elliptic and glabrous. Fruits are round,about 2.5 cm in diameter and very fragrant. Fruit peel thin andpulp soft with numerous seeds.

***P. Cattleianum* (Sabine) var. *lucidum* [Syn. *P. littorale* (Raddi) var. *littorale* (Berg.)]:** It is commonly known as lemon guava and relatively a hardy subtropical species. The fruitsare small, globose, juicy, acidic and sulphur yellow in colour.

***P.guineense* (Sw). [Syn. *P. molle* (Bertol), *P. araca* (Raddi), *P. Schiedeanum* Berg.]:** It is generally called as Brazilian jelly guava or Guinea or Castilian guava. Aslow growing shrubgrows 1-3m tall and it can withstand short periods of drought. The leaves are oblong, scantilyhairy on the upper side, coated beneath with pale orrusty hairs and distinctly dotted with glands. The fruits areround of 2-.5-3.0 cm diameter with yellow peel, pale yellow pulp surrounding thewhite central pulp and highly acidic. It contains numerous hard seeds (Mortan,1987). It is also reported to a source of wilt resistance (Edward and Gauri Shankar, 1964).

***P. friedrichsthalianum* (O. Berg) Medenza:** It is also known as Chinese guava or Costa Rican jelly guava.It is a tall tree about 7-10m high withslender and smooth branches. Leaves are oval or oblong, smooth,almost glossy above and pubescent below. Fruits are globose,sulphur yellow colour, small (2.5-3.0 cm) and sour. The fruits are good for jelly making becauseof their high acidity. It is reported to be wilt and *Meloidogyne incognita*nematoderesistant (Dinesh and Vasugi, 2010).It can be used as dwarfing rootstock too (Mishra and Goswami, 2016).

***P. cujavillus* (Burm.f.):** It is a shrub or small tree grows to a height of 1-4 m. Fruit is a globose berry, 2-3 cm diameter. It is cultivated in Indonesia and Philippines. This species is considered as a small form of guava. It can be used as a rootstock for guava (Mishra *et al.*, 2010) which improves vitamin C content in scion variety (Rajan and Markose, 2007) and it is also free from guava wilt (Singh *et al.*, 1977). It has shown tolerance to drought and sodic soils.

***P. montanum* (Swartz):** It is found in the mountains of Jamaica.Shoots are four angled, leaves oblong to oval,glabrous, fruits are round, pulp white with more number ofseeds. It produces fruits of poor quality.

***P. araucanum*:** It is grows as large tree with membranous leaves, brochidodromous venation, long petioles and peduncles, flowers solitary, axillary or ramiflorous or in short racemes, with two pairs of flowers. Fruits globose or pear shaped, thin pericarp, fruits yellowish green when mature, seeds angular or lenticulate.

***P. acutangulum* DC:** The shrub or tree ranges in height from 8-12 m. Its newly emerged shoots are quadrangular and winged near the leafbase and the new growth is finely hairy. Leaves are elliptical with very short petioles. Fruits are round to pear shaped,pale yellow to yellowish white acidic aromatic pulp containing few hard, triangular seeds.

Syzygium

There are about 400 to 500 species, of which a few provide edible fruits. Jamun (*Syzygium cuminii* Skeels) is an evergreen tree-attaining medium to large size. The trunk is thick and greyish white in colour. Leaves opposite, simple mostly firm and glossy, elliptic, pinnately veined with lateral veins close together; flowers in a few flowered panicles. Flowers are hermaphrodite, light yellow in colour, borne in the axils of leaves on branchlet; calyx lobes 4, calyx tube not extending beyond summit of ovary; petals 4, white, spreading; stamens many; ovary inferior, 2 celled (Mishra *et al.*, 2014). *Syzygium aromaticum* L. is grown for the sun-dried flower buds called cloves. Syzygium malaccense (Linn.) is known as Malay rose apple.

Feijoa

This genus is known for evergreen species viz., *Feijoa sellowiana* (Feijoa or pineapple guava. It is found wild in South America.

Varieties

Guava is largely a self-pollinated crop, but cross-pollination also does occur. This results in a large variability in the seedling population (Mishra *et al.*, 2018) from which promising genotypes have been selected in different agro-climatic regions of the country.

Allahabad Safeda: It is the most important variety of guava used for table as well as processing purposes. The tree is medium to tall in size, upright growth habit, heavy bearer, dense foliage and has a tendency to produce long shoots. Crown is broad and compact, often dome-shaped and rarely loose. Fruit is of medium size, round in shape with smooth skin and white flesh. The fruits are relatively soft with less seeds. Its keeping quality is good.

Sardar (Lucknow 49): Tree is vigorous, spreading and profuse bearer, heavy branching type with flat crown. Fruit is large, round to ovate in shape, primrose-yellow skin colour, white flesh and seeds are in plenty and harder than that of Allahabad Safeda.

Apple Colour: The trees are medium in vigour and are moderate yieldes. Fruits are medium sized, with apple coloured skin; cool temperature is required for good colour development. The pulp is white and firm, sweet to taste.

Lalit: Fruits are of saffron yellow colour with red "blush weighing 185-200 g fruif1. Its flesh is firm and pink with a good blend of sugar and acid. It is suitable for both table and processing purposes. The jelly made from this variety has better flavour and attractive appearance.

Sweta: The variety recently developed by Central Institute for Subtropical Horticulture, Lucknow is suitable for commercial cultivation. Tree is semi vigorous with medium height and is a prolific bearer. It is a variety with medium size globose fruits, creamy white skin with red spots or blush and snow-white flesh. Fruits are attractive and have good nutritive value.

Arka Mridula: At the Indian Institute of Horticultural Research, Bangalore, out of 200 open pollinated seedlings of the variety 'Allahabad Safeda' collected from Uttar Pradesh, one seedling selection, 'Selection-8', was found to be promising and was released as 'Arka Mridula'. The plants are semi-vigorous and spreading in nature. Fruits are round in shape and weigh about 180g. Fruit are yellow in colour with smooth skin. The pulp is white, firm, sweet with few soft seeds. The TSS is 12.0°B, 100 seed weight is 1.6g. Keeping quality is good. Pectin content is 1.04 %. The variety is good for jelly making.

Allahabad Surkha: It is an outstanding variety of large uniform pink fruit with deep pink flesh. The plants produce up to 120 kg fruits in the sixth year of fruiting. Trees are vigorous, dome-shaped and compact. The fruit is sweet and strongly flavoured with a few seeds. The plants are vigorous, dome shaped and compact. Trees are high yielding, producing upto 120kg per plant. The fruits are round with uniform, pink skin and deep pink pulp, sweet, strongly flavoured and with few seeds.

G. Vilas Pasand: The trees are vigorous, wide spreading with bushy, low growing habit. Fruits are round to ovoid, skin texture is course to smooth, fruit skin pale yellow to golden, colour of flesh is creamy white texture creamy soft, very large (400g to 800g) fruit, less seeds, very productive throughout the year. High content of Vitamin C makes it stand out among guava varieties.

Pant Prabhat: A selection was made as 'Pant Prabhat' for commercial cultivation at GBPUA&T, Pantnagar (Uttarakhand). Plant growth in this line is upright; with broad leaves, the tree is highly productive (100-125kg). Fruit skin is smooth and light yellow in colour, fruits medium sized with average fruit weight of 150-172g, pulp is white, seeds are small and medium soft, the fruit has a sweet taste with pleasant flavour and ascorbic acid content varies from 125mg in rainy season to 300mg/100g fruit weight in winter season. TSS varies from 10.5 to 13.5°B.

Dhareedar: A selection from an old, seedling orchard was made as 'Dhareedar' at the Fruit Research Station, Kuthulia, Rewa. The trees are vigorous, medium tall with erect and upright branching and a flat crown. Fruits are medium to large sized, roundish ovate in shape with 5-7 raised lines on the surface of mature fruits, the peel in greenish yellow, the pulp soft and sweet (TSS 11.7° Brix).

Chittidar: This variety is very popular in Western Uttar Pradesh. The fruits are characterized by numerous, red dots on skin. Fruits are sub globose, with white pulp, high TSS and Vitamin C content of 240 mg /100g pulp.

Hafsi: Fruits are spherical in shape with thin skin and medium size. The pulp is red with good taste and flavour. Seeds are comparatively less in number, but hard.

Red Fleshed: Fruits are medium sized with red pulp, round, smooth skinned, seeds are plenty and medium soft. Fruits possess sweet flavour, are rich in Vitamin C (386 mg /100g pulp).

Smooth Green: Fruits are round and medium sized. Skin is glossy and greenish yellow when mature. Pulp is white, good in taste and flavour.

Nasik: Fruits are medium sized, round, with white pulp, sweet with good flavour.

Banarsi Surkha: Trees are medium sized (5.1m) with a broad crown, fruit shape is round and surface smooth, skin colour golden yellow, pulp colour pink, seed number very high, seed texture very hard.

Seedless: The plants are very vigorous. There are different varieties, like, Saharanpur Seedless, Nagpur Seedless, Sringeri Seedless, etc., which are nearly identical. Two types of fruits, viz., long, big sized with warty surface, and yellow, thin skin with swollen calyx end and round; small, with very few seeds. The pulp is white, good to taste and has aroma, contains moderate to high levels of Vitamin C (240 mg/100g pulp).

Anakapalli: Fruits are medium sized with red pulp. Seeds are soft and plenty. Fruits are slightly oval.

Kohir Safeda: This is a cross cultivar of 'Kohir' and' Allahabad Safeda' Fruits ripen earlier, big in size with more amount of vitamin C and high yielding. It has been evolved at the Fruit Research Station, Sangureddy. It has been successfully grown in Rajasthan.

Safed jam: This is also a cross between Allahabad Safeda and Kohir. Fruits ripen earlier, big in size, few seeds in fruit, which are small and soft. It contains more amount of vitamin C and gives more yield than the parents.

Table 2: Guava cultivars grown in different parts of India

State	Cultivars
Andhra Pradesh	Allahabad Safeda, Sardar, Sweta, Lalit, Arka Mridula, Arka Amulya, Arka Kiran
Bihar & Jharkhand	Allahabad Safeda, Sardar, Apple Colour, Chittidar, Hafsi, Harijha
Uttar Pradesh	L-49, Allahabad Safeda, Apple Colour, Chittidar, Red Fleshed, Allahabad Surkha, Mirzapuri Seedless, Lalit, Sweta
West Bengal	L-49, Allahabad Safeda, Dudhe Khaja, Gole Khaja, Kabli, Baruipur ,Chittidar, Harijha
Karnataka	Allahabad Safeda, L-49, Arka Mridula, Arka Amulya, Arka Kiran, Navalur, Dharwar
Assam	Am Sophri, Madhuri Am, Safrior Payere, Allahabad Safeda
Maharashtra & Gujarat	Allahabad Safeda, L-49, Lalit, Sweta,Nagpur Seedless, Dharwar, Dholka
Madhya Pradesh	Allahabad Safeda, Sardar, Rewa-27, Gwalior-72, Lalit, Sweta

State	Cultivars
North Eastern States	Allahabad Safeda, Sardar, Red Fleshed
Tamilnadu	Anakapalli, Banarasi, Bangalore, Chittidar, Hafsi, Nagpur Seedless, Smooth Green

Attempt has already been made to find out suitable guava cultivars in red and laterite zone of West Bengal with 12 cultivars under rainfed condition. The average yields of Allahabad Safeda, Apple Colour, Lucknow-49 and Banarasi under rainfed condition at Jhargram were 29.2 kg, 34.2 kg, 35.9 kg and 32.4 kg plant^{-1} respectively while it was 71.6 kg, 69.5 kg, 42.3 kg and 73.7 kg plant^{-1} respectively under irrigated condition at Jhargram (Ghosh and Chattopadhyay, 1996). Another interesting result was also reported by them that the cultivars like Allahad Safeda, L 49, Apple Colour, Chittidar, Red Flesh and Arka Mridula showed better yield potentiality under agro-climatic conditions of Jhargram as compared to Ranchi, Jharkhand (Reddy *et al.*, 1999) or in Mandsaur, Madhya Pradesh (Mahour *et al.*, 2012). Introduction of promising guava cultivars in region may be an option for increasing the production and productivity. But the performance varied significantly with cultivars, location, agro-climate and soil type etc. The variation with regard to growth and bearing habits, yield, colour and quality among different guava cultivars were also reported by Chadha *et al.* (1981), Ojha *et al.* (1985), Deshmukh *et al.* (2013) and Singh *et al.* (2015) evaluated guava cultivars in varied agro-climatic conditions of the country. Singh *et al.* (2002) evaluated ten genotypes of guava under Sabour conditions of Bihar during 1998-1999 for their yield and quality attributes. 'Allahabad Safeda' was found superior to others with respect to yield (42.18 kg plant^{-1}) during winter season and fruit-quality attributes (12.00% TSS, 260.95 mg/100 g vitamin C during winter season) and it was at a par with 'Safed Jam'. Average number of flowers tree^{-1} was higher (812-561) during summer flowering (April-May) than that of rainy season flowering (July-August) in all evaluated genotypes, while fruit set percent was better during the rainy-season flowering (69.00-58.30%) compared with summer season flowering. The highest yield plant^{-1} (72.14-31.92 kg) was obtained in rainy season, followed by winter season (42.18-12.76 kg). However, maximum TSS (12.00-190.93%), total sugar (9.13-7.56), reducing sugar (5.69-4.20), vitamin C (260.95-201.00 mgl 100 g) and pectin content (1.42-1.04%) were found in the winter season cropping, followed by rainy season in all genotypes. Based on the yield and fruit quality attributes 'Safed Jam' and 'Allahabad Safeda' proved superior to others under Bihar conditions. Similarly, based on evaluation of 21 genotypes of guava, Jana *et al.* (2010) found the cultivars Allahabad Safeda, Sardar, Mild Fleshed, and Spear Acid Superior suitable for cultivation in the plateau region of Jharkhnd. The performance of different varieties/germplasm of guava was assessed under Malwa Plateau conditions of Madhya Pradesh,*viz.*, Dharidar, Chittidar, Apple colour, Rewa-72, Lucknow-49, Allahabad Safeda, Gwalior 27, Surkhi, Anakapalli, Godhliwala, Chaitpur, Nasik, Dharwar, Lal Guda, Smooth Green, Chitabuwala, Abuwala, Supreme Side Malta, China Red and China White. The maximum plant height was recorded in Dharidar (2.99 m) and the minimum in China White (1.43 m). The maximum fruit size (8.79 cm 8.44 cm) and weight (235.0 g) were observed in Allahabad Safeda while the minimum fruit size (3.57 cm 3.56 cm) and weight (31.0 g) were recorded in China Red. Total pulp content was maximum in Allahabad Safeda (97.74%) and minimum in China Red (85.91%). The seed content was found maximum in China red (14.09%). Besides, there were significant variation with regards to yield, vitamin-C, TSS and acidity in different genotypes. However, considering the overall performance of the different evaluated genotypes, it was recommended that Allahabad Safeda, Lucknow-49, Rewa-72, Dharidar, Apple Colour and Gwalior-27 are promising for cultivation in Malwa Plateau conditions of Madhya Pradesh (Mahour *et al.*, 2012).There are possible more than 400 cultivars around the world, but only a few dozen are responsible for the majority of plantings (Pommer and Murakami, 2009).

Ecology

Guava grows best in tropical and subtropical areas that are frost-free. Mature trees survive light frosts. It thrives well in both humid and dry climates from sea level to 1500 m. The optimum temperature required for guava cultivation and high yield of good quality fruit ranges from 20 to 30 C. Drought and low humidity during lowering can drastically reduce fruit set. Guava develops the best eating quality when they mature during dry period mostly in winter. The guava is a hardy plant that grows in most soil types. It responds well to soils with good drainage and high organic matter. Loam and alluvial types of soil is most ideal. Guava prefers a well-drained soil in the pH range of 5 to 7. Guava is fairly tolerant to soil salinity. While guava trees tolerate poor soils, fruit production is substantially enhanced when grown in rich soils under proper management (Mishra *et al.*, 2019).

Production Constraints

One of the major problems faced by the guava growers is the occurrence of wilt disease. This is a complex disease causing heavy losses to the growers caused by soil born fungi viz., *Fusarium solani, F. oxysporum psidi, Macrophomina phaseoli,* and *Rhizoctonia solani,* however, involvement of plant nematode is also reported (Goswami and Thakre, 2015).The source of resistance to guava wilt was reported by Edward and Shankar (1964), however, much research progress has not been made since then. Management of disease includes orchard sanitation and uprooting of diseased plants. Another problem is high number of seeds and seed hardness in diploid commercial varieties and poor yield with small, irregular and misshapen fruits of triploid seedless varieties (Negi and Rajan, 2007). Therefore, development of wilt resistant, high yielding less seeded guava cultivar deserves the first priority in guava breeding (Tiwari *et al.*, 2003, Mishra *et al.*, 2022). Poor shelf-life fruits and fruit quality are major problems faced by guava producers during marketing which can be managed by adoption proper nutrient management (Mishra *et al.*, 2003, Mishra *et al.*, 2003b, Goswami *et al.*, 2015 and Kumar *et al.*, 2017).In addition to this, anthracnose and canker cause severe losses, which can be managed by timely spraying of copper fungicides (Goswami and Thakre, 2005). The poor fruit set and the seasonal influence on the quality of fruits are the production problems which can be managed through proper canopy management and crop regulation practices adopting the horticultural techniques viz., high density planting system, proper training and pruning of trees, irrigation and nutrient management (Selvarajan*et al.*, 2005, Mishra, *et al.*, 2018, Mishra *et al.*, 2022).

Blossom Biology

The knowledge of flower bud development, anthesis, anther dehiscence and fruit set and extent of cross pollination are a pre-requisite for planned hybridization. In guava, flower buds are borne in leaf axil on current season's growth, either singly or two and three together on slender peduncles, pedicles about 2 cm long, bracts 2, linear. Calyx splitting irregularly into 2-4 lobes, whitish and sparsely hairy within; petals 4-5, white, linear-ovate 2 cm long, delicate; stamens numerous, filaments pale white, about 12 mm long, erect or spreading, anther straw coloured; ovary inferior, ovules numerous, style about 10 cm long, stigma green, capitate.. Guava is reported to require about 30-45 days from flower bud differentiation to complete development upto the calyx cracking stage. Three flowering seasons, viz., 'Ambe bahar', 'Mrig bahar' and 'Hatti or Hasta bahar' have been reported in different regions of India (Mishra et al.,2022).The flower buds that open the next morning show a cracking of the calyx on the previous day, nearly 24 hours in advance. In guava, peak anthesis observed to be between 6 and 7.30 A.M. under North Indian conditions (Singh and Sehgal, 1968). Dehiscence of anthers was observed to take place 15 to 30 minutes after anthesis and continued upto 2hrs (Balasubramanyam, 1959). Pollen fertility has been generally found to be high in guava (78 to 91%) in diploid varieties. Stigmatic receptivity was observed maximum on the same day of anthesis. The maximum fruit set occurs when the stigmas are pollinated within 2 hours after anthesis. For making crosses in a hybridisation programme, flowers are emasculated at least one hour before anthesis and bagged. It is pollinated within 2 hours after emasculation with freshly collected or stored pollens and rebagged. Bags are removed only after 5–6 days of pollination(Pommer and Murakami, 2009).

Genetic Resources

Guava, an allogamous fruit crop, exhibits out crossing to anextent of 35–40% (Nakasone and Paull (1998). Despitebeing an introduced crop in India, it exhibits a high degree of genetic variability probably owing to the continuous cultivation of heterogeneous seedling trees (Srivastava, 2005). Genetic diversity is critical for pooling genes andguava breeding for quality and stress tolerance. An accurate and refined assessment of available pool of genetic variation in production and quality attributes is, thus,desirable for the identification and selection of parents in a hybridization program for successful trait-breeding (Kherwar *et al.*, 2018). The broader objectives of guava breeding are aimed at improving both plant and fruit characteristics such as to develophig hyielding, high quality dwarf statured varieties with fruits of uniform shape, good size, attractive skin and pulp colour, fewer seeds and or soft seeds, resistant to wilt, long storage life, suitable for table and processing purposes and to evolve wilt resistant tomatoes warfing root stocks (Nimisha *et al.*, 2013)

Propagation by seeds during early days has given rise to considerable variability in growth habit, shape and size of the fruit, pulp texture, flavor and seediness. Guava germplasm is being maintained at several centers in the country in field banks which are often not systematically maintained. However, several guava germplasm of guava is conserved in the field genebanks of various SAUs and ICAR institutes in India Viz., CISH, Lucknow (152), IIHR, Bangalore (60), CHES (ICAR-CIAH), Godhra (70), FRS, Sangareddy (24), RFRS, Vengurlle (24), FRS, Rewa (42), GBPUAT, Pantnagar (18) and PAU, Ludhiana (13). In addition to this, guava germplasm is being maintained at CCSHAU, Hisar, BAU, Sabour, HETC, Basti and NDUAT, Faizabad. Some wild species of guava viz., *P. friedrichsthalianum* (O. Berg) Medenza, *P. cujavillus* (Burm.f.), *P. cattleianum*Sabine, *P. molle* Betrol and *P. acutangulum* DC were introduced from exotic sources which are being maintained in the field gene banks at CISH, Lucknow and IIHR, Bangalore (Mishra *et al.*, 2022).

Genetic Diversity Studies in Guava

Guava is self pollinated species with 35% cross pollination results in providing a heterogeneous, open pollinated seedling population with adequate genetic variation (Nakasone and Paul, 1998). Therefore, most of the guava varieties have evolved through selection from seedling population (Chaithanya *et al.*, 2014).Knowledge of the genetic diversity available and the origin of the cultivars would assist in the selection of parents for effective improvement programmes (Singh *et al.*, 2015). Surveys were undertaken by various workers to find out the diversity in guava fruit characteristics, and also to select elite seedlings from existing heterozygous seedling population having desirable horticultural traits.Seeds were collected from various guava growing regions and seedlings were evaluated to identify superior types for better yield characters. (Cheema and Deshmuk, 1972). Yadav and Shankar (2007) identified several elite seedling guava types based on bearing and fruit quality while surveying in Allahabad region. Similarly, Mishra et al. (2018) made an extensive survey of the diversity rich areas of Gujarat to earmark the elite guava genotypes having desirable traits in terms of shape, size, pulp colour, seed content, morphological and qualitative attributes of guava fruits. They observed twenty five genotypes of seedling origin for their quantitative and qualitative characters. Development of nutrient rich guava cultivars has been a focus of fruit breeding studies (Correa *et al.*, 2012) in Brazil and in India (Mishra *et al.*, 2022). The physico-chemical attributes of the fruit are important as high TSS and titrable acidity in fruits along with red pulp are desirable for processing industry and low acidity and high TSS are desirable for fresh consumption (Corrêa *et al.*, 2012).On basis of survey the studies showed that the genotypes collected from a private orchard was most prominent on the basis of its dwarfness and was found to be quite ideal as donor parent for further guava improvement in Punjab (Sandhu *et al.*, 2007).

Genetic distance matrix based on Jaccard's coefficient revealed maximum distance between Purple Guava and Allahabad Safeda (43%), whereas minimum distance was as low as 5.4% between two breeding lines HPSI-20 and HPSI-26. Interestingly half-sib progenies CISH-G-1 to CISH-G-6 had slightly more distance ranging from 10.8-24.0%. The clustering revealed that most of the cultivars/ accessions originated from Indo-Gangetic plains are grouped together (Bajpai *et al.*, 2008). Cluster analysis was also carried out using fifteen morphological characters in 29 varieties and 5 species at IIHR, Bangalore. In the first cluster, the species *P.cattleianum*, *P.friedrichsthalianum* and *P.molle* are placed. The second cluster consists of 8 varieties and one species, viz., Bangalore Local, Benaras, Dharwad, Karela, Kamsari, Spear acid, Surka Chitti, Surka Chitti Neputani and *P. quadrangularis*. In the third cluster, 10 varieties are grouped, viz., Chittidar, EC 147039, 147037, Hafsi, Nasik, Pati, Portugal, Sindh, Superior sour lucidum and White flesh. In the fourth cluster, Behat Coconut, Chakaiya Ruthmanagar, EC 147306, EC 147036, EC 147034, EC 162904, Florida seedling, G-6, L- 49, Mirzapur seedling, P. *chinensis* and Smooth Green are grouped together. The cluster means indicate that fruit weight, fruit volume, fruit length and width are greater more in Cluster II and low in Cluster I. Members of cluster I are mainly species that usually bear small sized fruits, except P. *quadrangularis*. The mean Vitamin 'C' content and fruit: seed ratio was maximum in cluster IV, which indicates that hybridization involving these accessions would be expected to result in maximum hybrid vigour. Mean acidity ranged over 1.0 to 1.65 % and total sugar was about 7.23 to 7.86g among different clusters. Crosses between Cluster I and Cluster II crosses may result in desirable combinations leading to development of varieties with good processing traits (Dinesh and Vasugi, 2010). A study was taken up during 2002-05 to generate information on the variability in different qualitative and quantitative parameters of the wild species of *Psidium*. The species evaluated exhibited wide variability with respect to growth and fruit quality attributes. The average fruit weight ranged from16.85 g in *P. friedrichsthalianum* to 157.2 g in *P. quadrangularis*. The flower diameter ranged from 1.6 cm in *P. friedrichsthalianum* to 4.3 cm in Lucknow 49. The species *P. friedrichsthalianum* may be used for dwarfness, as it is dwarf (2.7 m) compared with the other species. The high vitamin C content (231.5 mg/100g) in species like *P. chinensis* and profuse long duration of flowering (58 days) in *P. molle* could also be exploited in breeding (Vasugi and Dinesh, 2007).

High genetic diversity is also observed in population of guava in other parts of the world viz., Pakistan (Mehmood *et al.*, 2014), Venezuela (Aranguren *et al.*, 2010), Mexico (Padilla-Ramírez *et al.*, 2007), United States (Sitther *et al.*, 2014), Cuba (Rodríguez-Medina *et al.*, 2010) and Brazil (Correa *et al.*, 2012) considering the nutritional and economical importance of guava.

Breeding Techniques

Selection

Historically guavas have been grown from seed and seedlings were quite variable due to insect pollination of flowers. Seedlings segregate a lot which has been the basis of the variation used for selection all over the world by breeders as well as growers (Pommer and Murakami, 2009). In India, guava improvement work was initiated during 1907 at Ganeshkhind Fruit Experimental Station, Pune, primarily with collection of seeds of varieties grown in different places, to isolate superior strains. One strain from open pollinated seedlings of 'Allahabad Safeda' collected from Lucknow was selected and released as 'Lucknow-49' (Cheema *et al.*, 1954). Similarly, at CISH, Lucknow, seedling selections from half sib population of Apple colour guava, viz., CISH-G-3 (Lalit), CISH-G-4 (Sweta) have now become popular in many parts of country. Other important selected cultivars of guava are given below:

Arka Mridula: At the Indian Institute of Horticultural Research, Bangalore, out of 200 open pollinated seedlings of the variety 'Allahabad Safeda' collected from Uttar Pradesh, one seedling selection, 'Selection-8', was found to be promising and was released as 'Arka Mridula'.

Allahabad Surkha: At Allahabad Agricultural Institute, Allahabad, a selection from the local red pulp type has been released as 'Allahabad Surkha'. The fruits are round with uniform, pink skin and deep pink pulp, sweet, strongly flavoured and with few seeds.

Pant Prabhat: A selection was made as 'Pant Prabhat' for commercial cultivation at GBPUA&T, Pantnagar (Uttarakhand).

Dhareedar: A selection from an old, seedling orchard was made as 'Dhareedar' at the Fruit Research Station, Kuthulia, Rewa.

Hybridisation: In guava, flowers that are chosen for crossing are emasculated at the 'calyx rupture stage', a day before opening. Pollen from the pollen parent is brought from an unopened flower, preferably at calyx break stage. For making crosses, flowers are emasculated at least one hour before an thesis and bagged. It is pollinated within 2 hours after emasculation with freshly collected or stored pollens and bagged again. The stigmatic surface is gently smeared with the pollen during the morning hours between 08 am to 12.00 noon has given better results. Bags are removed only after 5–6 days of pollination.

Inter-varietal Hybridisation: Mostly inter-varietal crosses in guava are successful, having no crossability barriers. However, varietal cross incompatibility in guava is reported in crosses made between 'Behat Coconut' and 'Sardar', 'Behat Coconut' and 'Apple Colour' (Dinesh and Vasugi, 2010). The majority of existent breeding programs are under development in the countries where guava represents an important economic crop, for instance Brazil, India, Mexico, Pakistan, Cuba, Thailand and others. In India, breeding work for guava improvement has been going on at several research institutions and very few guava hybrids were released over the years and important one are listed below:

Safed Jam(Allahabad Safeda Kohir): Developed by FRS, Sangareddy. The tree is medium in height and a heavy yielder. Fruits are round in shape, large in size with a thin peel, good taste and few, soft seeds.

Kohir Safeda(Kohir Allahabad Safeda): The tree is vigorous, fairly large in size and dome shaped. Fruits are large, with few, soft seeds and white pulp.

Arka Amulya(Allahabad Safeda Triploid): Released by the IIHR, Bangalore. Plants are semi-vigorous and spreading type. Fruits are medium sized (180-200 g), weight of 100 seed is 1.8 g, pulp is white, with high TSS (12.5 B) and fruits have good keeping quality.

Arka Kiran(Kamsari Purple Local): Plants are semi vigorous, amenable to high density planting. The fruits are sub globose, weighing about 200-230 g The pulp is deep pink, thick and has good flavour. The seeds are medium soft (9.0 kg cm-2), with high lycopene content (7.45 mg/100 g) and TSS of 12.0 to 12.5 B.

Hisar Safeda(Allahabad Safeda Seedless): It is developed at CCSHAU, Hisar. Plants are upright, trees have a compact crown. Fruits are round, with a smooth urface, creamy yellow skin; average fruit weight is 92 g, creamy white pulpand few soft seeds with high TSS (13.4%).

Hisar Surkha(Apple Colour Banarasi Surkha): Fruits are round, skin yellow with red dots, average fruit weight 86 g, pulp pink, seed count medium, TSS high (13.6%).

Interspecific Hybridisation

Inter specific hybridization between *P. molle* and *P. guajava*has been tried and resultant hybrid plants are found resistant to wilt and are grafting compatible with commercial varieties of guava (Negi and Rajan, 2007). However, *P. guajava* and *P. molle* are cross incompatible when *P. guajava* is used as the female parent (Subramanyam and Iyer, 1982).

Heritability Studies

According to Narain (2010) heritability of particular trait, as a proportion of the phenotypic variation that is attributed to genetic causes is aprimeindicator found useful intaking decisions for genetic improvement of economic traits. Various genetic studies conducted in guava indicated that red pulp colour was dominant over white and that this character was governed monogenically. Many cultivated red fleshed varieties were found to be heterozygous for this character. Bold seeds were found to be dominant over soft seeds and this was also found to be determined monogenically. Linkage was also found between flesh colour and seed size, *i.e.*, red flesh with bold seeds (Subramanyam and Iyer, 1982). Mitra and Bose (1985) have reported heterosis in guava. Obovoid shape of fruit is dominant over round(oblate) and pyriform (Nimisha *et al.*, 2013).Studies conducted at Coimbatore by Raman *et al.*, (1969& 1971), have shown that triploidy and genetic factor(s) are responsible for female sterility and, that; variation among triploids is due to their independent origin from a distinct diploid variety. Iyer and Subramanyam (1971) opined that seedless varieties of guava were triploids, grew vigorously and bore fruits that were irregular in shape with ridges, because of irregular distribution of seeds of various sizes. Half-sib analysis in progenies of the variety 'Apple Colour' showed that genotypic variance was lower than the phenotypic variance, and heritability was moderately high for all the characters thus selection may be practiced for improvement of fruit characters and hybridization among less seeded diploids can be adopted in an improvement programme (Dinesh and Yadav, 1998). Inheritance pattern using 'Purple guava' and 'Arka Mridula' as parents that hybrids segregate in a ratio of 3:1 for green leaf types and purple leaf types (Dinesh and Vasugi, 2010).

Biotechnological Techniques

Biotechnological techniques are useful in breeding for disease resistance and in germplasm storage using tissue culture techniques. Use of *in vitro* culture and micro-propagation of superior guava cultivars has been discussed by several researchers (Mishra *et al.*, 2007). Jaiswal and Amin (1992) reported that somatic cell genetics could be useful in guava breeding with specific objectives. A technique for successful *in vitro* propagation using shoot-tip explants from mature trees was reported by Jaiswal and Amin (1987). These workers also demonstrated that adding activated charcoal to the medium enhanced rooting of explants and vegetative growth of established plantlets. Somaclonal variation could be useful technique for selecting guava plants resistant to the wilt disease. An efficient protocol for plant regeneration from callus culture would also be helpful in selecting plants resistant to disease or environmental stresses. Recovery of plants of haploid origin from anther/pollen culture of guava could offer advantages in breeding (Jaiswal and Amin, 1992).

Risterucci *et al*, (2005) constructed a library of microsatellite-enriched (GA)n and (GT)n and, 23 nuclear simple sequence repeat (SSR) loci were characterized in the guava. All the SSR loci were found to be polymorphic after screening for diversity in different cultivars, and across-taxa amplification tests showed potential transferability of most SSR markers in three other *Psidium* species. First to be published for *P. guajava*, this new SSR resource will be a powerful tool for genetic studies in guava, including cultivar identification and linkage mapping, as well as potentially, for, interspecific genetic studies within the genus *Psidium*. Nimisha et al. (2013) also advocated that traditional breeding procedures can help to a limited scale therefore, it is high time that biotechnological tools needs to be explored and exploited either alone orin combination with conventional breeding tools to improve the guava productivity and to address challenge of improving fruit quality, and to lerance to abiotic and biotics tresses.

Biotic Stress Tolerance

The most serious biotic stress faced the guava industry in the world are root knot nematode and guava wilt. At Horticulture Research Centre (MARDI), Kuala Lumpur, Malaysia68 accessions were screened for resistance to root knot nematode. Preliminary screening showed that accessions displayed potential resistance and one accession was identified to be resistant to root knot. These accessions are graft-compatible with local commercial clones of guava and thus could be used as a rootstock or parent for guava improvement programmes against the root knotnematode (Milan, 2010). Experimental results of González-Gaona*et al.* (2010) also provides evidence that guava genetic resistance to nematodes could be present in the Mexican guava germplasm under study, and could be used as rootstocks orhybridization in future works. Other studies have shown that higher levels of tolerance or resistance to root knot nematodes are present in wild *Psidium*species viz., *P. friedrichstalianum*, *P. longipes*, *P. cattleianum*or *P. arayan*. Some of these species have been evaluated as rootstocks in guava pointing out its compatibility and possible use (González-Gaona*et al.*, 2010).

Guava wilt, caused by soil born fungi is a serious disease of guava in the guava producing areas worldwide. Interspecific hybridization between *P. molle* and *P. guajava* has led to the development of hybrids which are resistant to wilt and are graft-compatible with commercial varieties of guava (Negi and Rajan, 2007). Schoeman and Labuschagne (2014) screened guava seedlings *in vitro*, they reported that selection MS44 showed some tolerance against the G2 isolate of the pathogen obtained from diseased 'TS-G2' trees, whilst selection MS70 showed some tolerance against the G1 isolate obtained from diseased TS-G1 trees. These selections were also resistant to the original Fan Retief isolate of the pathogen. Later, in vitro selection reported using *F. oxysporum* and *F. solani* culture filtrates for attaining resistance/tolerance to wilt disease in guava. *F. solani* (isolate F2 and F15) culture filtrate provided a positive selection agent in which susceptible shoots presumably grow slowly, and resistant shoots proliferate (Kamle *et al.*, 2012). However, second selection cycle allowed for the recovery of a higher percentage of resistant/tolerant plantlets. The source of variation, when linked to cell-based selection, can be used most effectively (Kamle and Baek, 2017).

Future Research Needs

Priority needs to be given to develop a wilt tolerant cultivar having high yield and fruit quality attributes like high TSS, good sugar-acid blend, aroma, attractive skin/pulp colour, soft seeds and processing traits like juice colour, high Vitamin C content, higher lycopene content, good pectin content and good flavour. Shelf life may be given adequate importance. In the recent past, efforts have been intensified to develop attractive red coloured cultivars for table and processing purpose. However, deep red colour has been found to be a very unstable character, hence, 'stable' types need to be identified and great care should be taken while selecting hybrids from a large population. Aneuploidy breeding should be intensified to develop high yielding, high quality varieties with fewer and soft seeds, and to develop dwarfing rootstocks. Induction of mutation by physical and chemical mutagens may be attempted where improvement in a specific character is required in an otherwise acceptable variety. Work on interspecific hybridization to develop wilt resistant rootstocks and scion varieties should be intensified. Varietal introduction, though, becomes an essential part of any crop improvement programme and needs to be made with great concern. Molecular characterization of germplasm needs to be accelerated with a view to work out genetical distance so that good recombinants can be arrived at by crossing suitable parents. Biotechnological tools need to be employed extensively. Attempts to spot genetic markers for wilt resistance may be made to improve efficiency for developing wilt resistant clones and rootstocks.

References

Agriculture and Food Agency. (2014). Index of agricultural production: guava. In: Agricultural Statistics Yearbook. Taipei: Council of Agriculture, p.87.

Aranguren, Y., Valecillos, C. and Fermin, G. (2010). Variability of Venezuelan Guava Geographic Landraces Employing Phenotypic Markers. *Acta Hort.*, 849:87-94.

Aular, J. and Casares, M. (2011). Consideraciones sobre la produccion de frutas en Venezuela. *Rev. Brasil. Fruticult.*, 33(1)187-198.

Bajpai, A., Chandra, R., Rajan, S. and Srivastava, N. (2008). RAPD and minisatelite markers for genetic diversity and relationship in guava varieties. *Indian J. Gen.& Pl. Breed.*, 68(4):441- 445.

Balasubramanyam, V.R. (1959). Studies on blossom biology of guava (*Psidium guajava* L.).*Ind. J. Hort.*, 16:69-75.

BBS-Bangladesh Bureau of Statistics. (2012). Statistical yearbook of Bangladesh. Bangladesh: Bureau of Statistics, Ministry of Planning.

BPS-Badan Statistica Indonesia. (2014). Sub-directorate of statistical Compilation and Publication. http://www.bps.go.id.

Chadha, K. L., Singh, H. P. and Tondon, B. K. (1981). A varietal trial of guava. In: Proc. Nat. Symp. Trop. and Subtrop. Fruit crop. 17.

Chaithanya, M.V.N., Dinesh, M.R., Vasugi, C., Reddy, D.C.L., Sailaja, D. and Aswath, C. (2014). Assessment of genetic diversity in guava (*Psidium guajava*) germplasm using microsatellites. *J. Hortl. Sci.*, 9(2):117-125.

Cheema, G.S., Bhat, S.S. and Naik, K.C. (1954). Commercial fruits of India. MacMillan & Co., New York, USA.

Cheema, G. S. and Deshmukh. S. B. (1972). Culture of guava and its improvement by selection inWestern India. *India Bull. Dept . Agric.Bombay*, p. 38.

Chundawat, B.S. (1990). Guava. *In*: Arid Fruit Culture. Oxford & IBH Publ. Co. Pvt. Ltd., New Delhi. pp. 123-135.

Corrêa, L.C., Santos, C.A.F. and Lima, G.P.P. (2012). Chemical and biochemical characterization of guava and araçá fruits from different regions of Brazil. *Acta Hort.*, 959:103-109.

DAFF-Department of Agriculture, Forestry and Fisheries. (2015). Local market fruit sales data. Pretoria: Directorate of Agriculture Statistics, <www.daff.gov.za>.

Deshmukh, N.A., Lyngdoh, P., Jha, A.K. Patel, R.K. and Deka, B.C. (2015). Comparative study on newly developed guava hybrids with commercial cultivars under mid-hills of NE India. *The Bios.*, 8(4): 1467-1470.

Dinesh, M.R. and Vasugi, C. (2010). Guava improvement in India and future needs.. *Hortl. Sci.*, 5 (2): 94-108.

Dinesh, M.R. and Yadav. I.S. 1998. Half-sib analysis in guava (*Psidium guajava*). *Ind. J. Hort.*, 55:20-22.

Edward, J.C. and Gaurishankar, G. (1964). Root-stock trial for guava. *Allahabad Fmg.*, 38:521-527.

El-Shereif, A.R. (2016). Tropical and subtropical fruits in Egypt. *ISHS Newsletter*, 14, Jan. 2016.

Ghosh, S.N. and Chattopadhyay, N.(1996). Performance of some guava cultivars under rainfed semi-arid region of West Bengal. *Hort. J.*, 9:121-27.

González-Gaona, E., Padilla-Ramírez, J.S., Lozano-Gutiérrez, J., España-Luna, M.P., Velásquez-Valle, R., Gallegos-Morales, G. and Cepeda-Siller, M. (2010). Evaluation of Mexican guava germplasm against root knot nematodes. *Acta Hort.*, 849:363-368.

Goswami, A.K., Lal, S., Thakare, M., Mishra, D.S. and Kumar, R. (2015). Studies on integrated nutrient management on yield and quality of guava cv. Pant Prabhat. *Indian J. Hort.*, 72(1):139-142.

Goswami, A.K. and Thakre, M. (2015). Management of production problems in guava. In: *Management of Production Problems of Horticultural Crops for Enhancing Productivity and Quality*. Singh, A.K., Singh, K., Prakash, J., Goswami, A.K. and Singh, S.K. (Eds.), Div. FHT, IARI, New Delhi, pp. 60-65.

Hayes, W.B. (1953). Fruit Growing in India. Kitabistan, Allahabad.

HORTGRO. (2015). Guava production and export data. Paarl: SATI, Information and Market Intelligence Division, <www.hortgro. co.za>.

Iyer, C.P.A. and Subramanyam, M.D. (1971). Problems with triploidy in guava. *SABRAO Newslett.*, **3**(1):31-33.

Jana, B.R., Munsi, P.S., Mgnna, D.C. and Das, B. (2010). Evaluation of guava (*Psidium guajava* L.) genotypes based on fruit morphology, physico-chemical properties and yield under eastern plateau condition. *Indian J. Plant Genet.Resour*, 23(I): 25-29.

Kamle, M. and Baek, H.K. (2017). Somatic embryogenesis in guava (Psidium guajava L.): current status and future perspectives. *Biotech*, 7:203 (DOI 10.1007/s13205-017-0844-0).

Kamle, M., Kalim, S., Bajpai, A., Chandra, R. and Kumar, R. (2012). *In vitro* selection for wilt resistance in guava (*Psidium guajava* L.) cv. Allahabad safeda. *Biotechnol.*, 11:163-171.

Kherwar, D., Usha, K., Mithra ,S. V. A. and Singh, B. (2018). Microsatellite (SSR) marker assisted assessment of population structure and genetic diversity for morpho-physiological traits in guava (*Psidium guajava* L.). *J. Pl. Biochem. Biotechnol.*, https://doi.org/10.1007/s13562-017-0438-2.

Kumar, J., Kumar, R., Rai, R., Mishra, D.S., Singh, S.K. and Nimbolkar, P.K. (2017). Influence of foliar application of mineral nutrients at different stages of guava. *J. Plant Nutr.*, 40(5):656-661.

Kumar, L.S.S. and Ranade, S.G. (1952). Autotriploid in guava (*Psidium guajava* L.). *Curr.Sci.*, 21:75-76.

Kwee, L.T. and Chong, K.K. (1990). Botany and Cultivars. In: Guava in Malaysia – Production, Pests and Diseases. Tropical Press, Kuala Lumpur, pp. 21–51.

Lucas, E.J., Belsham, S.R., Nic Lughadha, E.M., Orlovich, D.A., Sakuragui, C.M., Chase, M.W. and Wilson, P.G. (2005). Phylogenetic patterns in the fleshy-fruited Myrtaceae – preliminary molecular evidence. *Plant Syst. Evol.*, 251:35-51.

Mahour, M.K., Tiwari, R. and Baghel, B.S. (2012). Evaluation of guava varieties for growth, yield and quality attributes in Malwa Plateau of Madhya Pradesh. *Indian J. Hort.*, 69(4): 472-477.

MAM-Ministry of Agriculture Malaysia. (2013). Agriculture statistics handbook. Kuala Lumpur.

Martínez-De-Lara, J., Barrientos-Lara, M.C., Reyes-De Anda, A.C., Hernándezdelgado, S., Padilla-Ramírez, J.S. and Pérez, N.M. (2004). Diversidad fenotípica y genética en huertas de guayabo de calvillo, Aguascalientes. *Rev. Fitot. Mexi.*, 27(3):.243-249.

Mata, I. and Rodríguez-Mendoza, I. (1990). Cultivo y producción del guayabo. Editorial Trillas. México, D.F.

Mehmood, A. Jaskani, M.J., Khan, I.A., Ahmad, S., Ahmad, R., Luo, S. and Ahmad, N.M. (2014). Genetic diversity of Pakistani guava (*Psidium guajava*L.) germplasm and its implications for conservation and breeding. *Sci. Hortic.*, 172:221-232.

MNFSR. (2014). Ministry of National Food Security and Research. Government of Pakistan.

Medina, B. and Pagano, G. (2003). Caracterización de la pulpa de guayaba (*Psidiumguajava* L.) tipo "Criolla Roja". *Rev. Fac. Agron*-LUZ. 20:72-86.

Milan, A.R. (2010). Collection and evaluation of guava (*Psidium guajava* L.) for nematode resistance in Malaysia. *Acta Hort.*, 849:357-361.

Mishra, D.S., Berwal, M.K., Singh A., Singh, A.K., Appa Rao, V.V., Yadav, V. and Sharma, B.D. (2022). Phenotypic diversity for fruit quality traits and bioactive compounds in red-fleshed guava: Insights from multivariate analyses and machine learning algorithms. *South African Journal of Botany*, 149:591-603.

Mishra, D.S. and Goswami, A.K. (2016). High density planting in fruit crops. *HortFlo. Res. Spec.*, 5(3):261-264.

Mishra, D.S., Goswami, A.K., Kumar, R., Singh, S., Singh, A.K. and Singh, P.P. (2019). Guava. In: Origin and Biological Diversity of Horticultural Crops (Ed.) Peter, K.V. Brillion Publishing, Delhi, pp. 81-106.

Mishra,D.S., Misra, K.K. and Tiwari, J.P. (2003). Nutrient management in guava (*Psidium guajava* L.).*Adv. Hortic. Fores.*, 9:51-57.

Mishra, D.S., Singh, A., Kumar, R., Singh, S., Singh, A.K. and Swamy, G.S.K. (2014). Jamun. In: Tropical and Sub Tropical Fruit Crops: Crop Improvement and Varietal Wealth, Part.II (Ed.) Ghosh, S.N. Jaya Publishing House, Delhi, pp. 375-387.

Mishra, D.S. and Singh, A.K. (2022).Crop Regulation in Guava: A review. *Annals of Arid Zone* 61(2): 75-82.

Mishra, D.S., Singh, S., Singh, A.K., Apparao, V.V., Yadav, V. and Saroj, P.L. (2019). Crop regulation in fruit crops for improving quality and income of farmers. In: *Horticulture in Arid and semi-arid regions*, (Ed.) Jatav, M.K. NIPA, New Delhi, pp. 187-206.

Mishra, D.S., Singh, S., Singh, A.K., Yadav, V.Apparao, V.V. and Saroj, P.L. (2018). Assessment of genetic diversity in guava. *Indian J. Hort.*, 75(3): 362-368.

Mishra,D.S., Tiwari, J.P., Misra, K.K. and Lal, S. (2003). Effect of calcium nitrate and method of application on nutrient content of leaves and post harvest quality of guava *fruit.Scient. Hort.*, 8:11-19.

Mishra,D.S., Tiwari, J.P. and Lal, S. (2007). *In vitro* cloning of guava (*Psidium guajava* L.) cv. Pant Prabhat.Acta. Hort., 735:127-132.

Mishra, D.S., Tiwari, J.P., Kumar, R., Kumar, P. and Mishra, N.K. (2010). *In vitro* clonal propagation studies on the nodal explants of *Psidium cujavillis*. *Ann. Hort.*, 3(1):58-62.

Mitra, S. (2014).Tropical and subtropical fruits in India. *J. Trop. Crop Sci.*,1:1-2.

Mitra, S.K., Irenaeus, T.K.S., Gurung, M.R. and Pathak, P.K. (2012). Taxonomy and importance of *myrtaceae*. *Acta Hortic.*, 959: 23-34.

Morton, J. (1987). Guava. In: Fruits of warm climates. Eds. Julia F. Morton, Miami, FL. p. 356–363.

Nakasone, H.Y. and Paull, R.E. (1998). Guava. In: Tropical Fruits (H.Y. Nakasone and R.E. Paull Eds.), Wallingford, CAB International. pp. 149–172.

Narain, P. (2010). Statistical genomics and bioinformatics. *J. Hort. Sci.*, 5:85–93.

Nasution, F. and Hadiati, S. (2014). Characterization and clustering of some guava germplasm collections based on leaf and fruit characters. *Agrivita*, 36(1):91-96.

Negi, S.S., Rajan, S. (2007). Improvement of guava through breeding. *Acta Hortic.*, 735: 31–37.

NHB-National Horticulture Board. (2015). Indian horticulture database. Gurgaon: Ministry of Agriculture, Government of India.

Newsom, L.A. (1993). Native West Indian Plant Use. Ph.D. dissertation, University of Florida, Gainesville, University Microfilms, Ann Arbor.

Nimisha, S., Kherwar, D., Ajay, K.M., Singh, B. and Usha, K. (2013). Molecular breeding to improve guava (*Psidium guajava* L.): Current status and future prospective. *Sci. Hortic.*, 164:578-588.

Ocampo, F. and Núñez, V.M. (2007). Propagación in vitro de Psidium guajaba mediante organogénesis directa a partir de segmentos nodales. *Rev. Corpo.-Cie. Tecnol. Agro.*, 8(1):22-27.

Ojha, A.P., Tiwari, J.P. and Mishra, K.K. (1985). Studies on yield and quality of some guava cultivars. *Indian J. Hill Far.*, 4(1):15-18.

Padilla-Ramírez, J.S., González-Gaona, E., Perales de la Cruz, M.A., Gutierrez-Acosta, F. and Mayek-Pérez, N. (2007). Fruit Yield and Quality of Twelve outstanding selections of guava (*Psidium guajava*) from the Calvillo-Cañones region, México. *Acta Hort.*, 735:25-30.

Padilla-Ramirez, J.S., González, G.E., Esquivel, V.F., Mercado, S.E., Hernández, D.S. and Mayek, P.N. (2003). Caracterización de germoplasma sobresaliente de guayabo de la región Calvillo-Cañones, México. *Rev. Fitotec. Mex.*, 25:393-399.

Parra-Coronado, A. (2014). Maduración y comportamiento poscosecha de la guayaba (Psidium guajava L.). Una revisión. *Rev. Colomb. Ciê. Hort.*, 8(2):314-327.

Patiño, V.M. (1964). Plantas Cultivadas y Animales Domésticos en mérica Equinoccial. Tomo I: Frutales. Imprenta Departamental. Cali.

Pérez-Rodríguez, J., Arangurengonzález, M.; Luzbet-Pascual, R.; Reynaldo-Escobar, I.M.; Rodríguezexpósito, J. (2013).Aportes a la producción intensiva de plantas de guayabo (Psidium guajava l.) a partir de esquejes en los viveros comerciales. *CitriFrut*, La Habana, 30(2):11-16.

Pereira, F.M., Usman, M., Mayer, N.A., Nachtigal, J.C., Maphanga, O.M. and Willemse, S. (2016). Advances in guava propagation.*Rev. Bras. Frutic.*, 39(4):1-14 (e-358)DOI10.1590/0100-29452017.

Pereira, F.M. and Nachtigal, J.C. (2009). Melhoramento genético da goiabeira. In: NATALE, W.; ROZANE, D.E.; SOUZA, H.A.; AMORIM, D.A. Cultura da goiaba do plantio à comercialização. Jaboticabal: FCAV, Capeas, CNPq, FAPESP, Fundunesp, SBF, p.371-398.

Pommer, C.V. and Murakami, K.R.N. (2009). Breeding Guava (*Psidium guajava* L.). In: Breeding Plantation Tree Crops: Tropical Species (Eds. Jain, S.M., and Priyadarshan, P.), New York: Springer, vol.1, p p.83-119.

Pommer, C.V., Murakami, K.R. and Watlington, F. (2006). Goiaba no mundo. *Agronômico*, 58(1/2):22-26.

Rahim, M.A., Islam, K.K., Islam, M.S., Alam, M.S., Islam, F., Naher, N. (2011). Conservation of Guava (*Psidium guajava*) germplasm using wilt-resistant rootstock of Poly Piara (*Psidium cattleianum*). In: Roskaft, E. and Chivers, J. Present state, problems and prospects of its conservation. Bangladesh: University of Chittagong, p.127- 139.

Rajan, S. and Markose, B.L. (2007). Guava. In: Propagation of horticultural crops (Ed. Peter, K.V.). NIPA, N. Delhi, pp. 26-31.

Ray, P.K. (2002). Guava. In: Breeding Tropical and Subtropical Fruits. Narosa Publ. House, New Delhi, pp.143-154.

Raman, V.S., Rangaswamy, S and Manimekalai, F. 1971. Triploidy and Seedlessness in guava (*Psidium guajava* L). *Cytol.*, 36:392-399.

Raman, W.M., Manimekalai,G and Ramalingam, R.S. (1969). Observation on seedlessness, fruit development and cytology of varieties of guava. *Madras Agri. J.*, 56:255-261.

Risterucci, A.M., Duval, M.F., Rohde, W. and Billotte, N. (2005). Isolation and characterization of microsatellite loci from *Psidium guajava* L. Molec. Ecol. Not., 5:745-748.

Rodríguez-Medina, N.N., Valdés-Infante, J., González, G., Fuentes, V. and Cañizares, J. (2010). Genetic Resources of Guava (*Psidium guajava*) in Cuba: Germplasm Characterization and Breeding. *Acta Hort.*, 849:341-348.

Rodriguez-Medina, N.N., Valdés-Infante, J., Becker, D., Velázquez, B., Coto, O., Ritter, E. and Rohde, W. (2004). Morphological, agronomic and molecular characterization of Cuban accessions of guava (*Psidium guajava* L.). *J. Genet. & Breed.* 58:79-90.

Raseira, M., Do, C.B. and Raseira, A. (1996). Contribuição ao estudo do araçá. Embrapa – CPACT, Pelotas.

Reddy, N. N., Gangopadhyay, K.K., Rai, Mathura and Kumar, Ram. (1999). Evaluation of guava cultivars under rainfed sub-humid region of Chhotanagpur plateau. *Indian J. Hort.*, 56: 135-140.

Rodriguez, N.N., Valdés-Infante, J., Becker, D., Velázquez, B., Coto, O., Ritter, E. and Rohde, W. (2004). Morphological, agronomic and molecular characterization of Cuban accessions of guava (*Psidium guajava* L.). *J. Genet. & Breed.* 58:79-90.

Rodríguez-Medina, N.N., Fermin, G.A., Valdés-Infante , J., Velásquez, B., Rivero, D., Martínez , F., Rodríguez, J. and Rohde, W. (2010). Illustrated descriptors for guava (*Psidium guajava*). *Acta Hort.*, 849:103-109.

Sandhu, S., Khera, K.S. and Bal, J.S. (2007). Evaluation of guava germplasm (*Psidium guajava* L.) for vegetative and yield potential. *Haryana J. Hort Sci.*, 36(3/4):222-223.

Schoeman, M.H. and Labuschagne, N. (2014). Preliminary evaluation of guava selections for guava wilt disease resistance in South Africa. *South Afr. J. Pl. Soil*, 31(2): 109–112.

Selvarajan, M., Meenakshi, N. and Saraswathy, S. (2005). Production constraints and management in sapota and guava. In: *Management of production probles in Tropical fruits*. Balamohan, T.N., Vijaykumar, R.M., Soorianthansundaram, K., Poornima, K and Vadivel, E. (Eds.), HC&RI, Coimbatore, pp. 46-52.

Setiawan, B., Sulaeman, A., Giraud, D.W. and Driskell, J.A. (2001). Carotenoid content of selected Indonesian fruits. *J. Food. Compos. Anal.*, 14:169–176.

Singh, D., Gill, M.I.S., Boora, R.S. and Arora, N.K. (2015). Genetic diversity analysis in guava (*Psidiumguajava*) on the basis of morphological and physico-chemical traits. *Indian J. Agric. Sci.*, 85(5): 678-683.

Singh, G. (2008). High density and meadow orcharding of guava, CISH, Lucknow, India, 20 p.

Singh, R.B., Rostogi, S.S., Singh, R., Gosh, S. and Niaz, M.A. (1992). Effects of guava intake on serum total and high-density lipoprotein cholesterol levels and on systemic Blood pressure. *Am. J. Cardiol.*, 70:1287-1291.

Singh, S., Singh, J. and Hoda, M.N. (2002). Evaluation of guava germ plasm under Sabour (Bihar) conditions. *Indian J. Agric. Sci.*, 72 (7): 393-395.

Singh, U.R., Dhar, L. and Singh, G. (1977). Note on the performance of guava cultivars and *Psidium* species against wilt disease under natural field conditions. *Haryana J. Hort. Sci.*, 6(3-4):149-150.

Singh, R. and Sehgal, O.P. (1968), Studies on blossom biology of Psidium guajava L. (guava). II. Pollen studies, stigma receptivity, pollen and fruit set. *Ind. J. Hort.*, 25:52-59

Sitther, V., Zhang, D., Harris, D.L., Yadav, A.K., Zee, F.T., Meinhardt, L.W. and Dhekney, S.A. (2014). Genetic characterization of guava (*Psidium guajava* L.) germplasm in the United States using microsatellite markers *Gen. Res. Cr. Evol.*, 61:829–839.

Siwarungson, N., Ali, I. and Damsud, T. (2013). Comparative analysis of antioxidant and antimelanogenesis properties of three local guava (*Psidium guajava* L.) varieties of Thailand, via different extraction solvents. *J. Food Measur. Charac.*, 7(4):207-214.

Soubihe S.J. (1951). Estudos basicos para o melhoramento da goiabeira ´ (*Psidium guajava* L.). Sao Paulo: ESALQ. p. 166. Tese de Doutorado.

Srivastava, U. (2005). Genetic resources management in guava. In: Kishun R, Mishra AK, Singh, G., Chandra, R. (eds.) Proceedings of first international guava symposium. CISH, Lucknow, pp 17–18.

Subramanyam, M.D. and Iyer, C.P.A. (1982). Improvement of guava by breeding. Report, Fruit Workshop, Nagpur. pp. 117-118.

Tiwari, J.P., Bisen, B. and Mishra, D.S. (2003). Improvement of guava using indigenous strains. In: *Role of Indigenous Germplasm* In: *Improvement of Horticultural Crop*. Singh, R., Ram, H.H. and Tiwari, J.P. (eds.), Bishen SMPS Publ&Dis of Scientific Books, Dehra Dun, pp. 227-234.

Usman, M., Samad, W.A., Fatima, B. and Shah, M.H. (2013).Pollen parent enhances fruit size and quality in intervarietal crosses in guava (*Psidiumguajava*). *Int. J. Agric. Biol.*, 15(1): 125-129.

Valdés-Infante, J., Rodríguez, N.N., Velásquez, J.B.,Sourd, D.G., González, G.,Rodríguez, J.A. andRohde, W. (2012). Herramientas para un programa de mejoramiento genético del guayabo (*Psidium guajava*L.) en cuba. *Agron. Costarri.*, 36(2):111-129.

Varmudy, V. (2013). Guava: potential untapped. Facts for you. http://www. ffymag. com/admin/issuepdf/Guava FFYApril-13.pdf

Vasugi, C. And Dinesh, M.R. (2007). Genetic variability in some *Psidium* species. *Indian J. Agric. Sci.*, 77(7):420-423.

Yadav, L.P. and Shankar, G. (2007). Exploiting genetic diversity in guava (*Psidium guajava* L.) from Allahabad and surrounding areas. *Pt. Gen. Reso. News Lett.*, 149:14-16.

Chapter 8

Horticultural Technologies for Uttarakhand Himalaya

Rahul Dev, NK Hedau and Lakshmi Kant
ICAR-Vivekananda Parvatiya Krishi Anusandhan Sansthan
Almora (Uttarakhand)

The Indian Himalayas have an area of 53.7 Mha, or around 17% of the nation's total land area. Four of the country's 21 agroecological zones are wholly or partially encircled by the hilly agroecosystem. Between 150 and 4,000 mm of precipitation fell on average per year in the region. Throughout the year, the average temperature ranged between 8 and 22 °C. The cultivation phase of various crops varies annually from 90 to 270 days. There are 39 million people (4% of India's total population) living in the Himalayan region, agriculture provides the majority of the region's income (about 45%). Although certain states in the Himalayan region have even less agriculture than this, overall cropland in these states does not exceed 15% of the total land area. With more nuclear families entering the farming communities in the steep Himalayas, the amount of land available for cultivation has only decreased to 0.17 hectares per person. Climate conditions are changing, as are the resources that are available, particularly water. Unfortunately, due to the effects of urbanization, all of the good cropland in the Himalayan states' valleys—the only irrigated land in hills and mountains—is being converted to non-agricultural uses. The rising dominance of wild animals over farmers, which is currently increasing its disruptive value to mountain agriculture in all of the northwest Himalayan states, is the third biological cause of change after land and water.

Around 75–85% of the population of Uttarakhand depends on agriculture for their livelihood, with agriculture accounting for roughly 22.4% of the state's gross domestic product. As a result, the growth of agriculture and related activities is fundamentally tied to the development of mountain peoples. On November 9, 2000, the state of Uttarakhand was separated from the state of Uttar Pradesh in order to meet the high aspirations of the local population about development and higher living conditions. Yet, the topographical disparity between the plains and the hills inside Uttarakhand severely divides the state. In Uttarakhand, 86% of the land is covered by hill districts, although 86% of the state's farmland is on the Tarai plains. When it comes to income and way of life, the hill districts are less developed. A sizable fraction of the population suffers from malnutrition, and it has been shown that mountain regions have relatively greater rates of undernutrition than the rest of the country. According to research done in the Garhwal Mountains, almost 30% of the hill people are undernourished (Dutta and Pant, 2003). The adverse effects of the agroclimatic environment on the native population's nutritional state were made worse. The high rates of undernutrition in the mountains are attributed to dietary imbalances, the occurrence of diseases linked to nutrient deficiencies (such as protein, iodine, vitamins, calcium, or iron), poverty, high rates of maternal illiteracy, poor maternal & child health, and high morbidity rates. Moreover, the existing crop profile does not meet the daily nutritional needs of rural people in mountainous areas. The unpredictable nature of the weather defines Uttarakhand's climate. The yield of important cereal crops has already been impacted by unpredictable climatic events (dry spells/aridity, natural catastrophes). The implications of climate change on food output and farmers' livelihoods are thought to be quite substantial because of their over-dependence on seasonal rains. Due to their heavy reliance on climate-sensitive key crops, indigenous people living in mountains are now increasingly vulnerable and worried about their future. Due to changing climate trends including increased dry spells/aridity, natural disasters frequently seen in hill regions, and rising wild animal danger, the rural community is seeing a drop in the output of primary cereal crops. Families and a sizable portion of the population are relocating from the hill areas in search of alternate work and means of subsistence. Just before and after the creation of the state of Uttarakhand, migration rates have risen during the last few decades. Conditions have worsened in some hill districts to the extent that over 1700 villages have no one in there and many more are joining that category locally called "the ghost villages". Overall around 4000 villages are presently victims of migration. Consequent to migration, today, only less than 20% of agricultural land in the hill districts of Uttarakhand is being farmed, and the rest 80% has become fallow land for varying periods. The year 2020 has become a turning point as COVID-19 caused homecoming of migrants is expected to see part of the migrants abandoning their plans to migrate again and rather look for local opportunities to stay on their farmlands. We assume that there must be precious material that may have climate resilience, high productivity, and cash potential traits in the natural wild population of targeted horticultural crops. Promotion of these potential horticultural crops and their value addition for the development of ready-to-use products can be a strategy to sustain the livelihoods of rural communities and stop out-migration. Horticulture has its advantage in the Himalayan ecosystem due to its particular ecological condition and numerous micro-situations.

In the Himalayan states, the proportion of horticultural crops in mountain farmers' total agricultural income is rising. Farmers are abandoning or have already abandoned traditional grains like cereals, pulses, and native vegetables. The acreage and output of food grains in the Himalayas decreased even further from 2006–2007 to 2017–2018 due to a significant shift in cropping patterns away from grains and pulses and towards vegetables and fruits. As a result, these figures are now less than half of what they were in 2006–2007. According to Tulachan (2001), under horticulture (fruits and vegetables), ~ 13–15% is the gross cropped area in the Himalayan states. The climatic conditions in the Himalayas, range from subtropical (low-lying river valleys) to Alpine and cold (highly elevated regions) and are suited for diverse fruit cultivation. Besides this, the variety of fruit cultivation in this basin ranges from mango-guava, papaya, and citrus to apple, pear, walnut, kiwi, pomegranate, plums, peach, and apricot in the different elevations. Among the citrus fruit, lemon, elephant citrus, orange, malta, know, and others are prominent. In addition to this, vegetable (tomato, cauliflower, cabbage, beans, potato, capsicum, cucumber, and pumpkins) cultivation is also carried out on the slopes of the hilly basin at different altitudes where both indigenous tropicals, as well as temperate vegetables, are cultivated to a considerable extent. Among the flowering plants, orchids, tulips, anthuriums, gerberas, etc. can be cultivated in these regions. Tuber and rhizomatous crops like potato, tapioca, sweet potato, dioscorea, colocasia, ginger, and turmeric grow very well in these regions while the plantation crops like arecanut, coffee, and tea were introduced on a commercial scale in certain Himalayan regions. These potential horticultural crops are rich in nutritional and nutraceutical properties along with the ability to grow in harsher agro-climatic conditions. Moreover, their cultivation is also easy and does not require much care. Still, sufficient progress is not achieved so far, due to certain inhibiting factors, especially poor transportation, and marketing, that hinder the rapid growth of the horticulture sector in these hilly regions. The state of Himalayan farming highlights the fact that business as usual cannot continue. To meet this purpose, significant research, improvement, and development efforts are needed. For instance, high-value horticultural products dominate agriculture in the Himachal Pradesh state, accounting for over 44% of the cultivated land and roughly 48% of agricultural gross state domestic product (GSDP). This is mostly because to the variety of agroclimatic conditions, the ability to produce for "off-season" markets, the relatively high level of education of farmers, and the proximity of the production area to terminal consumer markets. In Himachal Pradesh, horticulture has contributed significantly to improvements in wages, employment, and, therefore, poverty alleviation. Notwithstanding the considerable opportunity for horticulture production in Himachal Pradesh to support targets for increased economic growth and poverty reduction. However, a success story in one state (Kashmir (apples) and Himachal Pradesh (vegetables and fruits), or one area in a state does not get repeated easily elsewhere. Proper land use planning with various horticulture and plantation crops must be done for the economic betterment of the rural people living in the U.K. Himalayan areas. If the state of Uttarakhand wants to achieve its potential and have an effect on sustainable development, it must solve a number of sectoral, institutional, and policy difficulties.

Indian Horticulture Scenario

Horticulture accounts for 18% of total cultivated land. In 2021-22, horticulture production will be 333.25 million MT from an area of 27.5 million hectares. Total Horticulture production in 2021-22 (2nd Adv. Est.) is estimated to be 341.63 Million Tonne, from an area of 27.74 million ha, an increase of approximately 7.03 Million Tonne (an increase of 2.10%) over 2020-21 (Final increases in the production of Fruits,

Vegetables, and Honey are anticipated, with a decrease in the production of Spices, Flowers, Aromatics & Medicinal Plants, and Plantation Crops).

Overall fruit yield and area are expected to increase to 102.48 million tonnes (2021-2022) from 102.48 million tonnes in 2020-21 and 7.01 million hectares (2021-22) from 6.93 million hectares in 2020-21. While the total area and production of vegetable crops increased from 10.85 Mh (2020-21) to 11.28 Mh (2021-22) and 200.44 MT (2020-21) to 200.46 million tonnes correspondingly (NBH, 2020-21). The area under horticulture has expanded by 17.80% in the last seven years, from 23.4 million ha to 27.74 million ha. Output has climbed from 280.9 million MT to 341.63 million ha (21.62%), surpassing India's food grain production (Fig. 1).

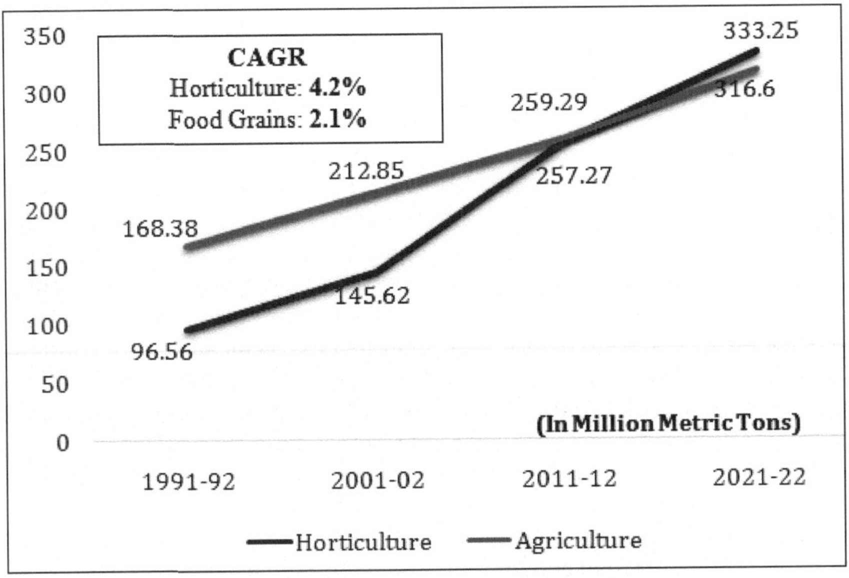

Fig. 1: Horticulture vis-à-vis Food grains production

Amongst vegetables, India ranks first in the production of ginger and okra and second in the production of Potatoes, Onions, Cauliflowers, Brinjal, and Cabbages. While India is also the largest producer of Bananas (26.45%), Mangoes, (including mangosteens and guavas) (43.80%), and Papayas (39.30%) (FAO, 2021). India is a net exporter of Horticulture produce of worth Rs. 45,963.24 Cr. Among different segments, the maximum contribution in the export was reported by spices (Rs. 29,273.86 Cr) followed by fresh fruit and vegetables (12,540.57 Cr) (APEDA, 2021). Grapes, Pomegranates, Mangoes, Bananas, and Oranges account for the majority of fruits shipped from India, while Onions, Mixed Vegetables, Potatoes, Tomatoes, and Green Chills make up the majority of vegetable exports. Fresh fruits and vegetables from India are mostly shipped to countries such as Bangladesh, the United Arab Emirates, Nepal, the Netherlands, Malaysia, Sri Lanka, the United Kingdom, Oman, and Qatar. Whereas processed fruits and vegetables are sent to the United States, the United Arab Emirates, China, the Netherlands, the United Kingdom, and Saudi Arabia (Tradestat, 2020-21). Despite India's small part of the world market (almost 1%), the country's horticultural products are becoming more acceptable. This is owing to parallel advances in the fields, capacity building of the farmer, processor, and exporter, development of cold chain infrastructure and quality assurance procedures, integrated post-harvest processing facilities across the country, etc.

The Horticulture Sector in Uttarakhand

On November 9, 2000, the state of Uttarakhand was formed out of the state of Uttar Pradesh. It is marked by different natural features ranging from snow-capped mountain peaks in the north to tropical rainforest in the south and is separated into Garhwal and Kumaon regions. Both regions are made up of 13 districts with a total geographical area of 5.35 million ha, of which 5.35 million ha (86%) is hilly and 0.74 million ha (14%). Forests cover the maximum land area (61%) and only about 14 percent of the geographical area is cultivable. About 53% of the cultivated area is rainfed and irrigated area is about 47 percent of the net area sown. Out of the total, about 89% are under small and sub-marginal. A vast number of people and land are subject to minor and marginal ownership. The soil in the tarai area is exceptionally rich and supports a variety of crops. The indiscriminate use of pesticides and overexploitation of groundwater make the soil in this region less fertile, resulting in a decrease in production sustainability (DoH and FP, 2018). The hill region, on the other hand, is prone to frequent soil erosion due to steep slopes, making it less and less fruitful, which may be avoided by using better management measures. Because of its geographical location and diverse climate, the state provides some distinct advantages for the development of horticulture, agro-processing enterprises, organic farming, off-season vegetable production, and medicinal and aromatic plant cultivation, all of which can be profitably utilised. The percentage of culturable wasteland is roughly 7%, signifying a substantial potential for fodder trees and other plantation crops like fruits.

Key Mountain Agriculture Types

1. Agriculture

Agriculture has a significant role in the state economy, accounting for around 23.4% of GDP. The majority of the state is covered by forests and wastelands, leaving just 7.41 lakh ha (approximately 14%) for agriculture out of a total stated area of 56.72 lakh ha (LUS, 2009-10). Rainfed agriculture is widely practiced throughout the state, particularly in steep areas. Irrigation facilities are generally accessible in plain areas and valleys in a few mountainous districts. Maize, rice, and pulses are dominant Kharif crops whereas wheat and barley are important Rabi crops. The average farm is low in the state i.e., 0.80 hectares, which has about 158% Cropping intensity. Food grain production increase varies greatly across plain and hilly parts of the state. As a result, the agricultural scenario portrays a contrasting image. Productivity in the plains of Udham Singh Nagar, Haridwar, Nainital, and Dehradun is high and similar to other agriculturally developed states in the country; however, productivity in the mountainous areas is quite poor, except in a few valleys. As a result, the development of the hill lags far behind.

2. Agroforestry

In the North-west Himalayas including Uttarakhand, agroforestry techniques commonly followed for centuries. Agroforestry supports the basic needs of fuel, wood, timber, and minor forest products of the farmers. In the hill region, agri-horticulture, Silvi-horticulture, horti-pastoral, and silvi-pastoral systems provide diverse land use alternatives for agroforestry. Agroforestry contributes to the socio-economic upliftment of hill communities while maintaining ecological equilibrium. A large farm animal population cannot be sustained only on the fodder generated by arable land. As a result, farmers must rely primarily on forest resources to support livestock; yet, during lean seasons (winter and summer months), farm tree fodder like *Grewia optiva, Quercus leucotrichophora, Quercus dilatata, Quercus semecarpifolia, Ficus nemoralis, Ficus palmata, Ficus roxburghii* can become significant fodder resources. Agroforestry can be a useful choice for meeting local needs for feed, wood, and firewood, reducing reliance on the forest. It can also help to prevent deforestation and erosion in the hills (Yadav et al., 2018).

3. Livestock

Around 90% of families in the state keep animals. Cattle farming is indigenous to Uttarakhand and is an essential component of the rural economy. It is an important industry that contributes to household income and family nutrition whereas also providing much-needed biomass and assisting various cultural and agronomic activities. Livestock is an essential part of millions of households across the state, and it employs poor and tenant farmers (Mittal et al., 2008).

4. Cash Crop Farming

Cash crop includes fruits, vegetables, medicinal and aromatic crops, honey, cut flowers suitable for improving the livelihood and economic status of the farmers. Important cash crops are potato, ginger and off-season vegetables. The state has the particular advantage of cultivating off-season vegetables in mountainous regions, which provides high market price to the producers (Bakshi, 2019).

5. Horticulture

Horticulture has become a vital component of Uttarakhand's economy. The state's geographical characteristics and climatic conditions are suited for the cultivation of subtropical and temperate fruit crops. Subtropical fruits such as litchi, mango, guava, aonla, pomegranate and citrus are produced on the state's plains. While temperate fruits like apple, pear, peach, plum, apricot, and walnut, are abundantly grown in this region. Potato, cauliflower, tomato, onion, brinjal, pea, cabbage, and okra are the most common vegetables cultivated in the state. Ginger, garlic, turmeric, and chilli are the main spices (DoH & FP, Uttarakhand; 2020-21). All horticulture components, such as pomology, olericulture, floriculture, spices, mushrooms, and medicinal and aromatic plants, have been well established in improving land productivity, generating employment, improving economic conditions of farmers and entrepreneurs, increasing exports, and, most importantly, providing nutritional security to the people. Data on the horticultural crop area, production, and productivity in the state are provided below.

Table 1: Area, production of fruit, vegetables and flowers in Indian Himalayas

Regions	Fruits		Vegetables		Flowers	
	Area (000'Ha)	Production (000' T)	Area (000'Ha)	Production (000' T)	Area (000'Ha)	Production (000' T)
NWH	755.042	3806.639	247.57	4095.19	2.47	68.93
NEH	467.922	4693.073	582.588	6611.868	5.709	136.893
Uttarakhand	178.8	670.63	100.14	1002.64	1.55	14.57

(National Horticulture Board, 2018-19)

Status of Horticulture in Uttarakhand

i) Fruits

Subtropical to temperate fruits are grown in the plain and hill regions of Uttarakhand. Some of the major fruits that are produced plain areas of state are Mangoes, Guava, Litchi, Citrus fruits, Anola, and temperate regions are Apricots, Oranges, Apples, Pear, Peach, Plum and Walnut. The major production districts in the state are Almora (1,59,101.0 metric tonne), Haridwar (1,04,460.50 metric tonne), Nainital (1,01,283.46 metric tonne), Pithoragarh (49,581.22 metric tonne) and U.S Nagar (47,674.05 metric tonne). On account of % share of area, Mango ranks first (36,598.39 ha) followed by apple (25,980.55 ha). Citrus obtains third position (2,175.3) whereas fourth and fifth place obtained by walnut (17,764.11 ha) and Pear (13,250.18 ha). In terms of production fruits like mango (1,50,666.45 metric tonne) followed by citrus fruits (86,851.56 metric tonne), pear (73,779.94 metric tonne), apple (64,879.26 metric tonne), and Apricot (25,621.01 metric tonne) (DoH & FP, Uttarakhand; 2020-21).

ii) Vegetables

Throughout the state, a wide variety of vegetables are grown, from subtropical to temperate varieties in plain areas to the snowline, respectively. Because most vegetables are short-duration crops, they suit quite well in the intensive cropping system and may provide extremely high yields with quite high economic returns to the producers. Major vegetables grown in Uttarakhand are Pea, French bean, coloured Capsicum, Onion, Tomato, Cauliflower, Cabbage, Cucumber, Potato, Garlic and Okra. In terms of production vegetables like green peas (54,925.50 metric tonne) followed by tomato (46,462.80 metric tonne), cabbage (32,756.39 metric tonne), Radish (26,063.04 metric tonne) and French bean (21,550.94 metric tonne). While, maximum area under vegetables was reported for Green Pea (6,917.13 ha), followed by Tomato (4,895.64 Ha), Cabbage (3,605.27 Ha), French bean (3,488.77 Ha) and Onion (2,752.46 Ha). The major district producing fresh vegetables are U.S. Nagar (98,256.03 Metric tonne) followed by Haridwar (91,760.10 Metric tonne), Pithoragarh (75,501.85 Metric tonne), Dehradun (74,193.80 Metric tonne), and Tehri Garhwal (64,139.11 Metric tonne). The area under vegetable cultivation was 72,958.46 Ha with a production of 6,56,628.07 metric tonnes in the year 2020-21 (DoH & FP, Uttarakhand; 2020-21).

iii) Flowers

Throughout the state, a wide variety of flowers ranged from loose to spike/cut flowers are grown. Most of the flowers being cash crops; capable of giving very high yields with very high economic returns to the growers. Major flowers that grown in state Gerbera, Carnation, Gladiolus, Lilium, Marigold, Rose, Tuberose Pea etc. The major district producing fresh loose flower are Haridwar (1,417.74 Metric tonne) followed by Dehradun (555.37 Metric tonne), U.S. Nagar (459.23 Metric tonne), Nainital (133.65 Metric tonne), and Pauri Garhwal (126.41 Metric tonne). Among different flowers, the maximum area is under, marigold (861.04 ha) which is followed by gladiolus (313.26 ha), rose (156.01), gerbera (108.49 ha), and carnation (18.56 ha). In terms of production. Marigold rank first with production of 2,681.87 metric tonne (MT) followed by gerbera (905.91 MT), gladiolus (466.17 MT), rose (173.53 MT) and carnation (52.37 MT) (DoH & FP, Uttarakhand; 2020-21).

iv) Spices

Mainly five spices grow in Uttarakhand are turmeric, chilli, coriander, garlic, and ginger. The total area under spices is 14,280.79 ha, of which the highest area is under ginger, which is 5,170.85 ha followed by chilli (2,751.03), garlic (1,924.23 ha), turmeric (1,790.81 ha) and coriander (1,390.48 ha). The total production of spices was 95,999.39 MT. The highest production was noted from ginger (50,045.42 MT) followed by turmeric (15,195.59 MT), garlic (11,270.19 MT), chili (9,333.48 MT), and coriander (3,931.89 MT) (DoH & FP, Uttarakhand; 2020-21).

Areas of Cultivation

The major share in area in the state is of fruits (62%), flowed by vegetables (35%), spices (9.44) and less area under flowers (1.55%). While, in production, vegetables (58%), fruits (39%), honey (2.8%), spices (2%) and flower (1%) occupies the largest share in State.

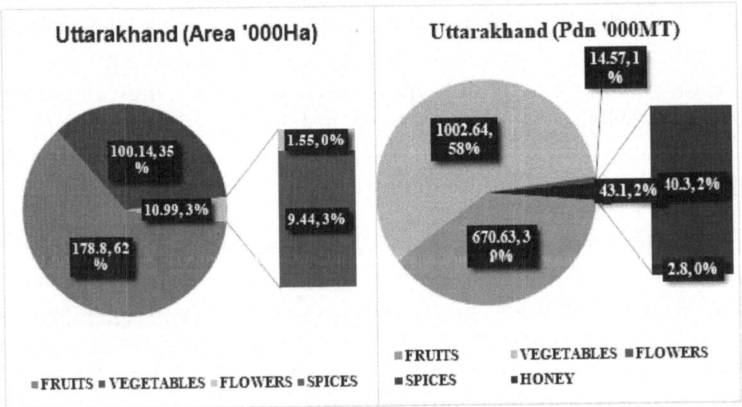

Fig. 2: Area and production of different horticultural components in Uttarakhand.

Table 2: Area, production of fruit, vegetables, flowers, and spices in different districts of Uttarakhand.

N	District	Fruits		Vegetables		Flowers			Spice	
		Area (ha)	Pro (MT)	Area (ha)	Pro (MT)	Area	Production		Area (ha)	Pro (MT)
							Loose Flower (MT)	Cut flower (lakh No)		
1-	Nainital	11033.20	101283.46	5914.90	60503.08	42.07	133.65	54.12	1442.54	8901.57
2-	US Nagar	7647.07	47674.05	8367.20	98256.03	137.85	459.23	210.73	1254.00	10250.40
3-	Almora	24324.00	159101.00	4646.00	43309.00	25.05	10.35	49.50	1350.00	9215.50
4-	Bageshwar	3703.99	12744.77	1578.19	7969.87	16.32	4.59	0.00	450.92	3547.10
5-	Pithoragarh	16210.61	49581.22	5700.23	75501.85	16.55	10.77	0.04	712.05	5580.77
6-	Champawat	8332.00	13778.00	3223.00	21455.50	8.35	2.54	14.64	776.00	4808.00
7-	Dehradun	27136.14	41643.20	11970.10	74193.80	218.55	55.37	90.42	1198.36	9785.14
8-	Pauri	21767.57	35795.91	5045.63	46853.24	36.60	126.41	7.00	942.28	6239.77
9-	Tehri	21123.61	28923.83	8268.76	64139.11	19.43	113.38	0.00	2171.12	16646.22
10-	Chamoli	4549.40	12832.50	2150.92	12336.76	62.10	15.50	0.00	571.67	2759.68
11-	Rudraprayag	3403.81	2594.00	1270.20	3560.73	80.60	78.46	10.65	642.20	2477.77
12-	Uttarkashi	15785.42	38483.44	10397.23	56789.00	143.10	94.91	3.49	1486.05	4869.97
13-	Haridwar	16054.00	104460.50	4426.10	91760.10	803.36	1417.74	1002.18	1283.60	10917.50
	Total	181070.82	648895.88	72958.46	656628.07	1609.93	3022.90	1442.77	14280.79	95999.39

Directorate of Horticulture & Food Processing, Uttarakhand (2020-21)

Opportunities for Mountain Horticulture Development

Uttarakhand has a natural variation in landscape and climate that is advantageous for crop diversification. As a result, alternative agricultural enterprise tactics such as fruit production, floriculture, and medicinal and aromatic plant production must be promoted. Moreover, less profitable crops can be replaced with more profitable horticultural crops. To increase system profitability, efforts should be made to incorporate vegetable crops into the cropping sequence. This technology has the potential to increase farmers' profitability by two to three times. Fruits, medicinal and aromatic plants, and other perennial horticulture crops can be grown on more land, especially if cultivable wastelands and fallow areas are used for this purpose. These areas could yield extra benefits if used for the production of perennial horticultural crops. At present, subsistence farming dominates agriculture in the higher altitude of the state, where only one season of cropping (Kharif) is feasible. In these very cold climatic conditions, mixed farming involving dairying, horticulture, agroforestry, and organic farming can be adopted for increasing farm productivity and rural livelihood. Among all, horticulture is a key factor in improving the income of marginal farmers of the mountainous region. It offers possibilities for diversification, greater employment, and a high rate of development over conventional agriculture. However, this faces a major challenge of low productivity. There was a substantial difference between the production and area share in horticultural crops. All horticultural subsectors, except spices, had very poor productivity. Despite low productivity, the floriculture, medicinal, and aromatic subsector in Uttarakhand is beginning to show potential. There is a tremendous opportunity to increase the yield of horticulture crops in Uttarakhand and turn this industry into a success. Some sector where proper planning and investment is required are as follow.

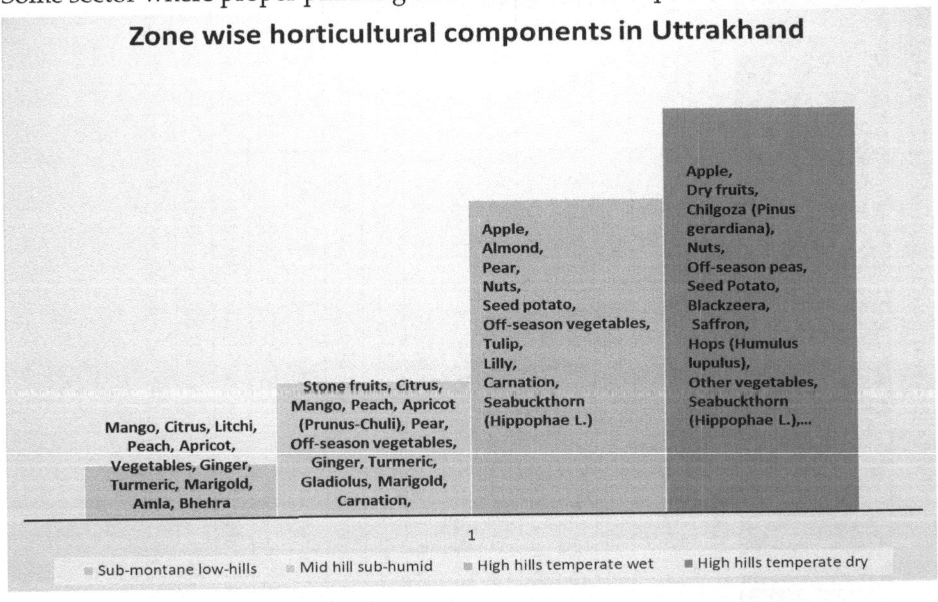

Fig. 3: Zone-wise horticultural components in Uttarakhand.

Flower and off-season vegetable cultivation

Hills are endowed with the ability to cultivate vegetables and other plants during seasons when they are not grown in plains. Although hills and hilly locations offer wonderful opportunities for farmers to plant off-season crops and a variety of flowers, the horticulture sectors are climate-sensitive (Bakshi, 2019). Farmers in the mid to upper hills have moved from apple agriculture to flower gardening or off-season food growing. Farming of flowers / cut flowers and off-season vegetables gives a comparative advantage to the farmers. Growing off-season crops such as tomato, peas, capsicum, french bean, cabbage, potatoes, carrot, sweet potato, and baby corn in mountainous terrain earns a fair market price, allowing these households to transition from food security to economic security.

Cultivation of medicinal and aromatic plants

The state has certain unique advantages for the cultivation of medicinal and aromatic plants and the development of oil extraction and distillation industries which can be gainfully exploited. There is huge commercial value in the oils and essences extracted from aromatic and medicinal plants. There is huge scope for the cultivation of medicinal plants like tej patta, amla, harad, kutki, stevia, atish, kutki, jatyamansi, chiraita, tagar, kuth, bhagera and aromatic plants like lemon grass, citronella, palmarosa, chamomilla, tulsi, geranium, naramotha, japanese mint, khas, and in the state and used extensively in the cosmetics industry. Because of the low production of aromatic and medicinal plants, oil extracted from plants is not linked with the pharmaceutical industry are produced, and buyers are not able to establish linkages (Mittal et., 2008). A positive attitude of farmers towards aromatic and medicinal plant cultivation will improve its production in the state. Intercropping of aromatic plants with food grains can also help diversify the income basket for small and marginal farmers

Development of a new cluster of high-value crops like Apple, Kiwi, Strawberry, and Papaya

Mountain agriculture is shifting away from conventional cereal crop farming and towards high-value cash crops such as apples (*Malus domestica*), almonds (*Prunus dulcis*), pear (*Pyrus communis*), peaches (*Prunus persica*), plums (*Prunus domestica*), and cherries (*Prunus cerasus*) and kiwi, strawberry, for local and export markets (Bakshi, 2019). The state's geographical characteristics and climatic conditions are suited for the cultivation of temperate and subtropical fruit crops. There is a need to develop cluster production and processing hub for targeted crops like apple, kiwi, papaya, and strawberry.

Table 3: Key producing Districts for the development of new clusters of horticultural crops.

SN	Crop	Key producing Districts
1	Tomato	Nainital, Udham Singh Nagar, Haridwar, Dehradun
2	Potato	Udham Singh Nagar, Almora, Tehri, Pithoragarh, Haridwar
3	Peas	Udham Singh Nagar, Tehri, Dehradun, Nainital, Haridwar, Almora
4	Pear	Almora, Pithoragarh, Nainital, Tehri, Pauri, Chamoli
5	Peach	Nainital, Almora, Pithoragarh, Chamoli
6	Plum	Almora, Nainital, Tehri, Pauri
7	Apple	Uttarkashi, Almora, Nainital, Dehradun, Chamoli
8	Herbal and medicinal plants	Almora (Rajwal, Dhali, and Jatti), Chamoli (Joshimath and Gopeshwar), Pithoragarh (Thal), Pauri Garhwal (Pokhal and Aita)

Agro-processing Industries

Small-scale agro-based industries and cottage industries should be promoted for the development of the hills. Increased production of fruits, vegetables, and aromatic & medicinal plants will bring in its wake the development of the horticulture-based preservation and processing industries. The oil extraction industry helps farmers to fetch good income by selling medicinal and aromatic plants and their byproducts.

Organic Farming

Almost 15,000 farmers and 45 crops are grown organically on 10,000 acres of land in Uttarakhand. There are various governments schemes nowadays. There are incentive programmes available to encourage farmers to switch to organic farming. There is still a need to promote and organise the state's dispersed organic activities and initiatives for organic horticulture. Farmers in the state may be likely to adopt organic farming due to the prevalence of food grains-based rain-fed mixed farming in the hills. Also, practically all organic products have a higher market price than conventional farmed items.

Challenges of the Horticulture Sector

The horticulture sector has emerged as a potential sector in the State economy. An increase in demand for horticultural produce due to greater health awareness, rising income, export demand, and increasing population poses the challenge of further increasing the production and productivity of horticultural crops. There are certain inhibiting factors. The lack of connectivity from villages to the road and the market, lack of good cold chain storage and transport networks, less mechanization and low productivity, shortage of inputs, and lack of marketing are other major challenge the farmers face that hinders the rapid growth of the horticulture sector in these hilly regions (Mittal et al., 2008). Additionally, land use degradation and climate change has arisen as a serious environmental issue and resulted in increased temperature, more demand for water, and an increase in abiotic and biotic stress, which has a direct impact on agriculture and horticulture, especially productivity (Janakiram and Reddy, 2016). To meet the above challenges, we need innovations for improving horticultural crop production in the changed scenario.

The Main Reason for Low Productivity

Fruits:

- Most of the orchards of apple, litchi, and mango, citrus are senile and unproductive
- Lack of High Yield Varieties/ quality planting materials
- Need for capacity building & technological Interventions
- Area under micro irrigation
- Low-density plantation
- Lack of adoption of global best practices
- Marginal farmers
- Fragile and less land holding

Vegetables:

- Poor irrigation facilities (only 10% irrigated).
- Low area under micro irrigation.
- Non-availability of quality seed.
- Seed production chain is neglected.
- Major share of vegetable seed comes from private sectors and they do not concentrate on hills due to low seed volume as compared to plains.
- Low area under Protected cultivation.
- Lack of adoption of global best practices.
- Marketing is another problem, especially in perishable vegetables.

Approach to Increase the Productivity of Horticultural Crops

- True to type, disease-free, healthy planting material suitable for subtropical and temperate climatic conditions.
- Rejuvenation of old senile orchard of Apple, Litchi, Citrus, Mango, and Peach
- Pollination Support through Bee Keeping: Horticulture Mechanization
- Human Resource Development (HRD)
- Post-Harvest Management (PHM), Marketing, and Processing Infrastructure
- Anti-bird/hail nets: protection of fruits from hailstones, birds, etc.
- Creation of water resources: Community tanks, Water harvesting takes and links to drip or sprinkler irrigation system

- Protected Cultivation: construction of new poly houses, Shad nets,
- Adaptation of High Yield Varieties
- Need for capacity building & technological Interventions
- Adoption of global best practices

Table 4: Productivity of Horticultural crops in Himalayan State

SL	Himalayan States	Productivity (q/h)			
		Fruits	Vegetables	Flowers	Spices
1	Arunachal Pradesh	02.61	06.64	-	05.93
2	Assam	15.07	12.53	17.93	02.99
3	Himachal Pradesh	02.48	20.11	35.80	02.55
4	Jammu & Kashmir	07.42	22.24	112.32	00.25
5	Manipur	09.65	07.79	05.80	02.49
6	Meghalaya	09.28	10.50	02.00	04.81
7	Mizoram	05.39	04.73	13.88	03.65
8	Nagaland	09.29	11.09	610.00	05.90
9	Sikkim	02.84	05.96	69.13	02.49
10	Tripura	10.34	17.41	-	04.91
11	Uttarakhand	03.75	10.01	09.40	04.27

Strategies to Meet Challenges in the Horticulture Sector

The Green Revolution benefited the farming system on the state's plains while neglecting the mountainous region. Besides from the dangers, there are several potential for boosting output and productivity, particularly in horticultural crops, cultivable wasteland availability, and rainwater gathering activity conservation. There is also a good chance for organic horticulture/farming, agricultural diversification through the horticulture component, post-harvest technology, market intervention strengthening, and farm machinery utilisation to make horticulture a more profitable employment. The diversification in the agricultural sector mainly of the horticulture component has become a major source of positive growth for the sector itself and for the state. Horticulture is a most sensitive sector because of climate change and is the main occupation of the rural communities in the hills so the changing climatic conditions have a severe impact on these communities, although hills and mountainous regions offer a great opportunity for farmers to cultivate off-season vegetables and a variety of flowers (Mittal et al. 2008). As an organised industry, the horticulture sector with strong forward and backward links may generate and maintain growth. The main strategies for meeting this challenge will be the development of new cultivars of horticultural crops that are tolerant to high temperatures, resistant to pests and diseases, have a short duration, and produce a high yield under stress conditions, as well as the adoption of high-tech horticulture and prudent natural resource management. Because a substantial portion of the state is covered by hills, different sources of irrigation are required to raise the net irrigated area, which in turn will enhance the state's cropping intensity. These alternatives include rainwater gathering, check dams, hydrams for lift irrigation, and so forth. Drip irrigation, sprinklers, and other water-saving technologies can also be utilised to improve water management. If properly encouraged, the horticulture industry can emerge as a viable source of revenue acceleration, job creation, poverty reduction, and export promotion. Some of the specific suggestions can be applied to each district for the better harnessing of potential of specific climatic conditions (Mittal et al., 2008).

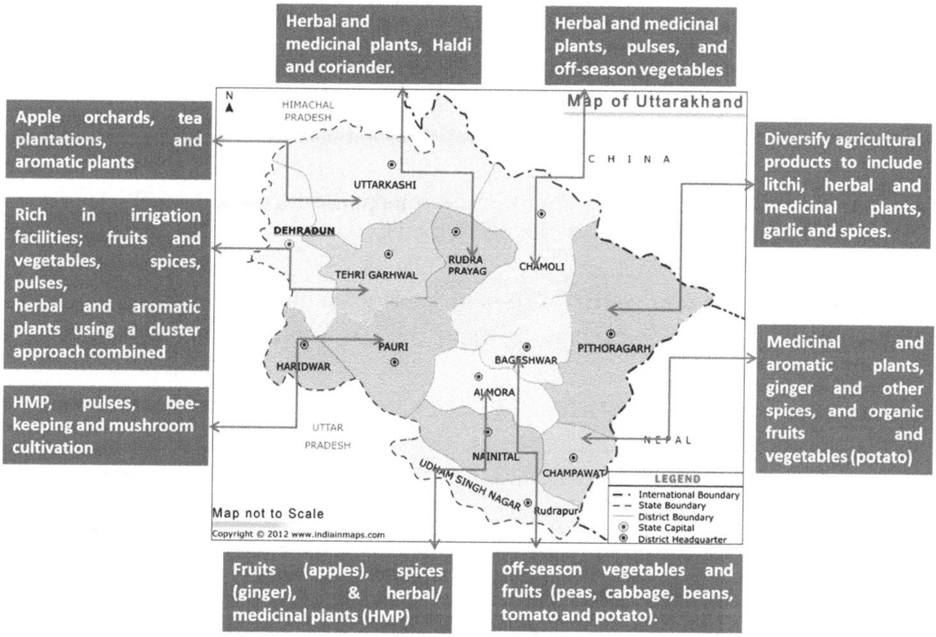

Fig. 4: Development Strategy for Hill Districts of Uttarakhand.

Conclusion

Hill Horticulture has some inherent constraints of remoteness, inaccessibility, marginality, land fragility, short cropping season, and lack of entrepreneurship. All these have led to the under-utilization of resource bases in the hills and a limited generation of surpluses. A new generation of farmers: the educated unemployed youth exploring entrepreneurship opportunities and yet to benefit from agricultural diversification through different components of horticulture. They need to be equipped with the necessary knowledge and skill in scientific fruit/ vegetable/ flower/ medicinal and aromatic plant production, entrepreneurship, and agribusiness for better income.

References

Agricultural Produce Processing and Export Development Authority (APEDA) (2021):Product profile of horticulture. Url; https://agriexchange.apeda.gov.in/product_profile/prd_profile.aspx?categorycode=0101# accessed on 19.12.2022.

Bakshi, P. (2019). Mountain agriculture development in India. pp 173-270.

Department of Horticulture & Food Processing (2018) State Profile: State Horticulture Mission, Government of Uttarakhand. url: State Profile: Department Of Horticulture, Government Of Uttarakhand, India (Uk.Gov.In) accessed on 18.12.2022.

Department of Horticulture & Food Processing (2020-21). District wise horticulture crops area and production statistics. Directorate of Horticulture & Food Processing Chaubatia, Almora, Uttarakhand.

Dutta, A. and Pant, K., (2003). The nutritional status of indigenous people in the Garhwal Himalayas, India. *Mountain Research and Development*, 23(3), pp.278-283.

FAO. (2021). World Food and Agriculture – Statistical Yearbook 2022. Rome. https://doi.org/10.4060/cc2211en

Janakiram, T and Reddy, V.R. (2016). Hill horticulture – status, problems and strategies. Journal of Hill Agriculture 7(1): 1 – 11.

Mittal S, Tripathi G and Sethi D. (2008). Development Strategy for the Hill Districts of Uttarakhand; Working Paper No. 217. Indian Council for Research on International Economic Relations.pp.

National Horticultural Board (NHB). (2018). Indian Horticulture Database.

National Horticultural Board (NHB). (2020-21, 1st Adv. Estimate). Indian Horticulture Database.

National Horticultural Board (NHB). (2021-2022, 2nd Adv. Estimate). Indian Horticulture Database.

Tradestat, Ministry of Commerce and Industry (2021-22): https://tradestat.commerce.gov.in/eidb/ecom2.asp

Tulachan PM (2001) The state of mountain agriculture in India. In: Rana GM, Shrestha B, Joshi SM (eds) State of mountain agriculture in the Hindu Kush-Himalayas: a regional comparative analysis. ICIMOD, Kathmandu, pp 25–31

Yadav, R. P., Bisht, J. K., Meena, V. S., & Choudhary, M. (2018). Sustainable Agroecosystems for Livelihood Security in Indian Himalayas. Sustainability of Agroecosystems. doi: 10.5772/intechopen.74495

Chapter 9

Utilization of Exotic Varieties in Mango Improvement

Ashish Yadav*, Vishambhar Dayal, Vineet Singh and S. Rajan
ICAR-Central Institute for Subtropical Horticulture,
Kakori, Lucknow, (U.P.)
Email: 2005ash@gmail.com

Mango is the most significant fruit crop of India. Millions of farmers are dependent on mango for their livelihood security. India conquers first place in mango production of the world and accounts for almost half of the global production and area. India exports mango to over 40 countries worldwide both in fresh and processed form. Details of India's exports of mangoes are given in Table 1. As per the latest data, the output of mango stood at 20.26 million tonne during the 2019-20 and is estimated to increase 4.24 % to 21.12 million tonnes in the year 2020-21.

Table 1: India's exports of mangoes {Quantity (in MT); Value (in Million USD)}

COMMODITY DESCRIPTION	2015-16		2016-17		2017-18		2018-19	
	Quantity	Value	Quantity	Value	Quantity	Value	Quantity	Value
Mangoes Fresh	36779.26	50.10	52761.00	66.46	49671.32	59.45	43063.70	53.80
Mangoes Sliced Dried	3644.09	6.94	376.78	2.40	569.05	3.05	1035.56	3.39
Mango Pulp	128866.01	121.29	130886.07	126.14	110923.73	104.54	72187.96	64.12
Total	169289.36	178.33	184023.85	195.00	161164.10	167.04	116287.22	121.31

*Source: DGCI&S, *Figures of 2018-19 are provisional only*

Exotic Mango Varieties

Mango genetic resources play an important role in developing newer varieties as per the farmers need and it will become more vital in near future. In mango, so far natural genetic variability has been exploited for improving productivity as well as introducing new varieties with specific traits. In this context, maintenance of exotic mango gene pool is important for current and future breeding programs. At ICAR-CISH, collection of mango variety started in 1975 when the ICAR-CISH started as Central Mango Research Station, Lucknow. Kensington from Australia was first exotic variety introduced at Central Mango Research Station, Lucknow, then Tommy Atkins and Sensation were added from USA. At ICAR-CISH, currently 18 exotic mango varieties are being collected and conserved from various countries. The field gene bank has not only conserved the exotic mangoes varieties but also made them available for breeding purpose for developing new mango varieties. The brief description of some important exotic mango varieties is given in Table 2.

Table 2: Brief Description of Exotic Mango Varieties

Exotic Mango Varieties	Parentage	Origin	Remarks
Haden	Mulgoba x Haden seedling population selection	Florida, USA	Haden is having superior quality fruit, brilliant color and good flavor. Earlier, Haden was one of the most widely cultivated in the world, however, gradually its acreage reduced largely due to fungus problems, along with inconsistent production, and problems with internal breakdown of the fruit. Most of the mango varieties developed in Florida were either direct or indirect descendants of Haden.
Tommy Atkins	Haden x Unknown	Florida, USA	Fruit possesses good commercial potential due to its color, strong and steady production, relatively good fungus resistance, very long shelf life and good handling characteristics. It is the main mango comprising about 80% of mangoes sold in the United Kingdom and United States.
Sensation	Haden x Brooks	Florida, USA	Fruit skin colour is dark plum red, flesh is light yellow with very fine fiber and flavor is mildly sweet with a light aroma, high yielding, late-season mango cultivar, problems with uneven ripening. The tree has good production characteristics. It is a vigorous grower, developing an open canopy.

Exotic Mango Varieties	Parentage	Origin	Remarks
Edward	Haden x Carabao	Florida, USA	Vigorous tree with large and dense canopy, moderately resistant to anthracnose, and is a regular but poor bearer. Fruit is oval/oblong shaped with an undulating surface; skin is bright yellow in color with a pink to reddish tint with small white speckles. Flesh is tender, fiberless, juicy, and deep yellow to orange in color with sweet flavor, pleasant aroma and outstanding eating qualities.
Eldon	Cowasaji Patel x Unknown	Florida, USA	The fruit is oval in shape having variegated color upon maturity, and can be a mix of green, yellow, orange and red blush. The flesh is yellow and has a sweet flavor with a pleasant aroma. The fruit contains a monoembryonic seed. The trees are moderately vigorous with a large canopy that contains light green leaves.
Irwin	Seedling of the Lippens cultivar that was open-cross pollinated with Haden	Florida, USA	Irwin gained commercial acceptance due to its good production, relative disease resistance, and attractive color. Fruit skin develops an eye-catching dark red blush at maturity. The flesh is yellow, fibreless and has a mild but sweet flavor and pleasant aroma. Tree is moderately vigorous.
Keitt	Brooks x Unknown	Florida, USA	Fruits are large size; skin color is typically green with some light red blush. Flesh has fibers, tangy and sweet, with monoembryonic seed. The fruit generally has good disease resistance and late-season ripening.
Kent	Haden x Brooks	Florida, USA	Kent is popular in Florida. The fruit is of oval shape, and has a rich, sweet flavor, excellent taste and lack of fiber. It will usually turn a greenish-yellow color with some red blush as it matures. The seed is monoembryonic. It is susceptible to anthracnose, and the fruit's poor shelf life limited its commercial scale in Florida.

Exotic Mango Varieties	Parentage	Origin	Remarks
Osteen	Haden x Unknown	Florida, USA	Osteen is grown on commercial scale due to its color, production characteristics, and flavor. The smooth skin has a yellow background color but usually turns dark purple. The flesh is nearly fiberless and has a mild yet sweet flavor. Fruit contains monoembryonic seed.
Palmer	Haden x Unknown	Florida, USA	'Palmer' mango is a large, commercially grown late-season mango cultivar. Fruit skin is yellow with red blush when ripe; flesh is orange-yellow and has a mild and aromatic flavor, with minimal fiber, and contains a mono embryonic seed.
Kensington Pride	Selected from *a seedling in Bowen, North Queensland in the* '1870s.	Australia	It is Australia's most popular mango, accounting for >80% of the country's annual commercial mango market. It is considered to have a distinctive flavour and aroma with polyembryonic seed. Fruit skin color is yellow, developing some red blush. The flesh is yellow, with moderate fiber, and has a sweet and spicy flavor.
Maya	*Maya* variety was discovered as a natural clone in an Israeli orchard.	Israel	*Maya* is a high quality mango with a lovely flavour and smooth, fibreless flesh. The shape is quite round compared to many mangoes, and a distinctive feature is a small prominent 'horn' at the stylar end. The peel colour is a deep yellow, with an attractive orange-red blush. The peel is thin and the stone quite small. The seed is monoembryonic.
Starch	-	Trinidad	Creamy, rich, very sweet, yellow pulp, lots of fiber, very susceptible to sap burn and anthracnose on the tree and when picked ripe. Shelf life is short (3-5 days). Very productive. Vigorous upright grower. Gets mango seed borer insects a lot in Trinidad.

Exotic Mango Varieties	Parentage	Origin	Remarks
Cowsaji Patel	-	India	Cowasjee Patel (1744–1799), became the Patel of Bombay in 1763. A variety of mango, grown in his orchards in Thane and Poona, was named after him. The mango, "Kawasji – Patel", was gathered green with white pulp. It had no fibre and hence was an excellent choice for cooking. The Europeans and the Parsees preferred the mango to cook jams and jellies. Mango plants sent from the Ganeshkhind Botanical Gardens still stand tall at various farms and gardens in Europe and the US. These include the "Kawasji – Patel" variety too.

Utilization of Exotic Mango Varieties

The mango improvement program in India started in the beginning of 19th Century. Afterwards the mango improvement work has been taken up by several research stations in the India especially Sabour and Kodur in early forties. During 1950, mango improvement work was started in Punjab and in 1951 at Saharanpur to develop regular bearing varieties. At IARI mango improvement program was started in 1961. Till 1980, the major focus was on the development of mango varieties for high yield, regular bearing, dwarfiness for its suitability in high density plantation, good fruit quality and long shelf-life of fruit for domestic market. After 1980s, in the mango improvement program additional objective was defined for the development of red-peeled varieties with lesser sweetness, good sugar: acid blend and long shelf-life of fruit for export markets utilizing exotic mango varieties as a parent. The mango hybrids developed by using exotic mango varieties as a parent are given in Table 3.

Table 3: Utilization of Exotic Mango Varieties in the Development of Mango Hybrids

Released Hybrids	Parents	Year of Release	Released Institute	Important Characteristics
Pusa Arunima	Amrapali x Sensation	2002	IARI, Pusa, New Delhi	Regular bearer, semi-vigorous and suitable for closer planting (6 m x 6 m). Fruits mature by first week of July. Fruits are medium to large (230 to 250 g), attractive red blush on the shoulders of fruits, medium TSS (19.5 Brix) and long shelf-life (10-12 days). It is suitable for both domestic and international markets, with long shelf-life (10 to 12 days) at room temperature after ripening.

Released Hybrids	Parents	Year of Release	Released Institute	Important Characteristics
Pusa Surya	Selection from Eldon (EC No. 141457)	2002	IARI, Pusa, New Delhi	Trees semi-dwarf and suitable for closer planting (6m x 6m). Fruit ripens by mid-July in northern India, medium to large (260 to 290 g) with attractive apricot-yellow peel colour with pink-blush. Pulp with medium TSS (19.0 Brix) with long shelf-life (10 to 12 days) at room temperature after ripening. Suitable for both domestic and international markets.
Pusa Pratibha	Amrapali x Sensation	2011	IARI, Pusa, New Delhi	It is regular bearing variety with attractive fruit shape, bright red peel and orange pulp. It has oblong shaped, uniform sized fruits. The plants are semi-vigorous. It has good sugar: acid blend and uniform fruits. It has 7 to 8 days shelf- life at room temperature after ripening. The plants are semi-vigorous and about 278 plants of this hybrid may be accommodated in a hectare (6 m x 6 m) as against 100 plants of Dashehari (10 m x 10 m). On per plant basis, it yields about 3.0 times higher than Dashehari, which includes 'On' and 'Off' year fruiting. It has good sugar: acid blend and above all uniformity in fruit size, which is lacking in Amrapali.
Pusa Shrestha	Amrapali x Sensation	2011	IARI, Pusa, New Delhi	Plants are semi-vigorous and suitable for closer planting (6m x 6m). Regular bearing with elongated fruits and attractive red peel. Pulp is orange in colour, fibreless and firm at ripening, moderate sugar: acid blend with uniform fruit size (228g) and higher pulp content (71.9%). It contains good amount Beta -carotene and ascorbic acid. The total soluble solids are 20.3%, vitamin C (40.3mg/100g pulp) and ß-carotene content (10,964 ug/100g pulp). It has pleasant flavour with shelf life (7 to 8 days) at room temperature. It is suitable for domestic market as well as international market.

Released Hybrids	Parents	Year of Release	Released Institute	Important Characteristics
Pusa Lalima	Dashehari x Sensation	2011	IARI, Pusa, New Delhi	The plants are semi-vigorous and are suitable for closer planting, regular bearer, attractive oblong fruits with bright red peel on yellowish green background with orange pulp and good sugar : acid blend. The fruit (209 g) having attractive red peel colour and higher pulp content (70.1%). Pulp with medium total soluble solids (19.7%), vitamin C (34.7 mg/100 g pulp) and high ß-carotene content (13,028 µg/100 g pulp). It has good flavour with shelf life (5 to 6 days) at room temperature. It is suitable for domestic market as well as international market.
Pusa Deep-shikha	Amrapali x Sensation	2020	IARI, Pusa, New Delhi	It is a hybrid having regularity in bearing and uniform sized fruits. Fruit are oblong in shape, bright red peel with orange-yellow pulp, moderate TSS (18.67%), high pulp content (70%) and ascorbic acid content (35.34 mg/100 g pulp), β-carotene content (9.48 mg per kg pulp), good shelf-life (7 to 8 days) at room temperature. It is semi-dwarf and suitable for medium density planting (6 m x 6 m). On per tree basis, it yields 50.33 kg (~10 year- old trees) with estimated yield of 14 t/ha.
Konkan Samrat	Alphonso x Tommy Atkins	2014	RFRS, Vengurle, Dapoli, Maharashtra	Medium fruit size (284.50 g) as compared to the Alphonso (250 g) and Tommy Atkins (484 g). The pulp percentage (73.28) is similar to Alphonso (74.60 %) and Tommy Atkins (71.79 %). Good T.S.S. (20.03 ^{0}Brix). This variety has high percentage of perfect flowers (27.52 %). Regular bearing, cluster bearing, spongy tissue free and low fibre. This is second hybrid having Alphonso as female parent and first hybrid of exotic parent. This variety is ideal for table purpose.

Compared to outstanding Indian cultivars, the Florida mango cultivars are highly adaptable to many agro-ecological regions and are regular bearer. These cultivars are highly attractive and develop red blush at maturity, firm flesh, a high flesh to seed ratio and regular bearing habit are reasons for their increasing popularity. Florida cultivars, *e.g.*, 'Tommy Atkins', 'Keitt', etc. are also moderately resistant to anthracnose, an added advantage during prolonged storage and export. ICAR-CISH has about 35 years old Mango Breeding Program. Various hybrids are being developed using exotic mango varieties as a parent e.g. H-1042, H-1723, H-1739, H-2047, H-2709, H-3669, H-3803, H-3842, H-4015, H-4061, H-4065, H-4189, H-4208, H-4252, H-4267 and H-4352. Some of them are promising type and are in the advance stage of evaluations.

The physical characteristics of mango fruits like fruit weight, size, per cent of fibre content etc. play a crucial role in selection for table purpose and processing type mango varieties. The high pulp percentage is an ideal parameter from the consumer's point of view. Mango has an established domestic market and possess bright opportunities for export in the international market, whether in fresh or processed forms.

References

Bhujanga Rao, C. and Rangacharlu, V.S. 1958. Breeding new mango varieties in South India. *Indian J. Hort.*, 15:173-83.

Sen, P.K., Mallik, P.C., Ganguly, B.D. 1946. Hybridization of mango. *Indian J. Hort.* 4:4.

Singh, R.N. 1996. Mango. New Delhi. ICAR.

Singh, R.N.; Sharma, D.K. and Majumder, P.K. 1977. Improvement of mango through hybridization. In: Fruit Breeding in India. G.S. Nijar (Ed). Oxford and IBH Publishing Co., New Delhi.

Wagle, P.V. 1929. *Agri. J. India*, 24: 254-263.

Chapter 10

Cosmic Farming in Horticulture The Concept and Way Forward

R. K. Pathak

Homa Teacher and Preacher, Fivefold Path Mission, International Organization
Former Director, ICAR-CISH, Lucknow (UP)
President, Society of Horticultural Research & Development, Uttarakhand

Looking adverse impacts of Industrial Farming, Bhartiya Kisan Sangh, in collaboration of concern ministry of Government of India, state government, institutions, NGO's and few select people organized conferences separately on Panch Mahabhoot and at the end a week conference from 20-26th February on the name of Sumangalam Panchmahbhoot was organized at Siddhagiri Math Kolhapur, Maharashtra. It concluded that enormous quantum of infinite source of energy of Panch Mahabhoot i.e., Cosmos is available in the country.There is dire need to involve few institutions to conceive techniques through which, a part of the energy can be mediated in crop production for sound health and harmonious survival at planet Earth. Introspection of statements is indication for revamping the current farming practices and even look ancient practices conceived by Gurukul's of the country to save the humanity. I do feel satisfied that it's because of Almighty grace, after getting feedback from resources constrained small and marginal farmers of the country, who shared that they are unable to bear costs of farming, are forced to move distant cities along with family just for survival, leaving elderly parents, few cows and birds in the villages. This has forced my inner conscious to conceive alternate technology for survival of this vulnerable group in the native villages. From over one and half decades ofmy field experience, I propose 'Cows based cosmic farming' as viable alternative.This will be able to boost all production factors incrops in all situations, simultaneously addressing issues of basic elements of mother nature.

This technology is ideally suited to resource constrained small and marginal farmers, farm women who can nurture few cows and use their products.

Cosmic Farming in Horticultural Crops

Cosmic Energy is the fundamental source of energy to everyone including plants. The Vedic answer is that cosmic energy is behind all mysteries of nature. It is available through primitive microorganisms in the soil as also in the plant cells. Therefore, plants have capacity to produce all food elements needed for their growth, without feeding them specific elements.

Horticultural crops are grown for high production, nutritive, therapeutic and aesthetic values. Owing to varying size, suitability to different climate, crops are suited for production in variable production conditions i.e., commercial, kitchen, backyard, roof top, indoor house, front of house, pool, different kinds of structures, multi-storeyed, even over the huts and nearby by compost pits and many others ways. Its commercial production and local consumption will have for reaching impacts on human health and harmonious life.

There is minimum gap between farms to fork in most of the crops. Often pesticides are used in evening and crops are harvested in morning. Few are even dipped in chemicals for giving shining appearance. It is pertinent to mention that, in recent decades, trend for consumption of organic commodities is expanding rapidly in different countries including India. Few are eaten in raw forms or for extraction of juice, if contaminated with cocktail of agrochemicals, may act as potent source of health hazards. Looking strength of country "Cosmic production and consumption" of horticultural foods will have for many positive implications in resolving number of crises being faced by the humanity.

On deep introspection, it is evident that Mother Nature is very much benevolent to humanity,has made elaborate arrangement to humans for harmonious survival at the planet. The five nature's representatives are: Cow – Dung- Urine- Soil microbial consortia including dwellers and vegetations. By integration of these cosmic energy can be mediated in crop production of horticultural crops.

Nature's Gesture to Mediate Cosmic Energy

- Soil microbial consortia, soil dwellers including earthworms have capacity to mediate cosmic energy.
- There are number of biological N-fixture (BNF), including leguminous crops which are capable of mediation of free N available in the biosphere.
- Millions of lips beneath foliage are busy in devouring carbon dioxide and exhaling oxygen essential for survival of human being and other living one.
- Chlorophyll in foliage mediates solar energy in process of photosynthesis, the basic step in preparation of food needed for survival of everyone.
- Without plants we cannot breathe, nor drink and not eat.

- Cow with hump which is pyramid shape, belly is mini replica of cosmos and horns have capacity to mediate cosmic energy.
- Belly of indigenous cow, udder and teats are rich in beneficial microbial consortia, hence the dung, urine and milk obtained from these cows are immense use in agriculture & human health.

Farming basedon systematic and synergistic mediation of infinite source of energy of cosmos, is being promoted 'Cosmic Farming'.

I admit that, its because of Grace of Almighty, I was able to conceive cheap, effective, affordable techniquies, effective in mediation of energy of Panchbhoot.

Techniques for Mediation of Cosmic Energy

- Practice of use of BD calender for different farming activities.
- Habitat development through intensive plantation of different kinds of plants including, those belonging to legumunous group.
- Management of few cows with hump and use of their produtcs.
- Enhancing soil fertility through crop rotation, inclusion of legume as sole crop, cover, inter, green manures, pump crops as per feasibility.
- Practice of mulching whenever, whereever possible as thumb rule.
- Enhancing soil humus formation and its maintenance.
- Assertive promotion of homa therapy and use of Agnihotra ash.

Use of BD calendar: Life pattern of living organisms are interwoven into cosmic rhythms. Human life, animal and plant life are strongly dependent on rhythms of the cosmos. Sun is ultimate source of energy but moon light is essential for the growth of the plants. On the basis of such influences, planting calendar is prepared for the year. The agricultural operations such as compost preparation, use, field preparation, seed sowing / transplanting, harvesting, storage performed during different timings of the year are better impacts. Besides, BD-preparations are prepared and used as per direction of calendar. Its non-monetary input, but have many advantages in ensuring quality production.

Habitat Development: Proper habitat development in a given place is essential which is done by encouraging wide range of plants, trees, bushes, annuals, creation of water bodies, enhancing soil humus content. Plants act as source of biomass; provide congenial conditions for birds, predators, parasites, restrict soil erosion, aerate the soil and thus helpful in cosmic farming. A large number of plants are Biological Nitrogen Fixers (BNF) and potent source for mediation of atmospheric N (78.6%) freely available.

Soil: Soil fertility is important, critical, and crucial for survival of humanity at the planet earth. It is only fertile soil, rich in organic carbon (humus) content can assure quality production, healthy animals, plant health and ultimately the human being. Unfortunately,most of human activities directly or indirectly are adversely impacting the three basic elements of nature that is the soil, water, and nearby air, hence soilfertility and crop productivity. Health of soil is of utmost importance for feeding an ever-growing population at planet earth. It seems that we are now heading towards such a situation of soil collapsing. According to information from FAO (Food and Agriculture Organization of the United Nations), now about one third of the world's soil has already been degraded. In fact, long ago ancient literature has clearly cautioned for respecting mother earth.

Quote from ancient literature, BC-3500 years back- Upanishad's

"Upon this handful of soil, our survival depends. Nurture properly, it will grow our food, fuel, shelter and surround us with beauty. Abuse it, the soil will collapse and die, taking humanity with it."

Somehow due to misconception, Earth is being treated as dumping ground by many human activities thereby, indicating her disapproval through un- imageable climatic disasters in one or parts of the world.

This is most opportune time to reorient our ancient inheritance and see its feasibly in the present context, what and how the planet 'Earth' can assure healthy, happy & harmonious life to future generations.

The soil itself can be viewed as a living organism, because it is a habitat for plants, animals and micro-organisms all are interlinked with each other. The microbial consortia are potent source to mediate cosmic energy. Mobilization of elements from deeper horizon to upper soil is possible through deep penetrative roots of plants and active participation of earthworms and other many organisms. Humus in soil helps in regulated supply of nutrients and moisture to plants mediates moisture from dew and cosmic energy from atmosphere. Quality humus formation occurs after proper decomposition of organic biomass duly supplanted with animal wastes. It is biologically rich soil can provide all nutrients and moisture as per need of the plant.

Mulching- is integral component in cosmic farming and need to be practiced wherever, whenever possible. It plays major role in maintenance of soil fertility by minimizing soil erosion, water requirement, weeds management. Regular practice of mulching will keep the soil cool during winter, warm in summer, prevent water loss through evaporation. It provides optimal situations by management of soil temperature, moisture, and dark conditions, friable soil for root development, enhancement of earthworm and termite activities thereby soil fertility. This in turn will enhance other soil life indispensable for plants. The mulch ultimately decomposes and enriches the soil. Plays major role in maintenance of soil fertility and enhancing quality production with minimum expenditure.

- Helpful in saving water requirement through minimization of runoff, transpiration and cutting solar radiation.
- Reduces overheating and cooling of soil and thus helpful in maintaining continued root and soil biological activities.
- Encourages development of white fibrous roots, real organ for absorption of plant nutrition from soil and cytokinin's production.
- The fibrous roots grow in upper layer of soil and hence it is essential to

keep upper layer or top few inches in friable condition.

- With many vegetables and some fruits direct contact with soil is avoided.
- Continuous mulching and its drenching with bio enhancers, enhances organic carbon in the soil and its maintenance.

Things to note- *We change our cloth with change in weather conditions, but plants are exposed to natural vagaries. Mulching provides a cushion and assure conducive conditions equipped to fight off pests, diseases and water stresses for soil bio life thereby its remarkable impact on quality production.*

Once the soil gets better aeration through these processes, giving them better access to subsoil moisture, and additionally, capillary action is improved. When plants can develop their full root system, their metabolism is strengthened and they are much better equipped to derive nutrition from soil and atmosphere for quality production.

Earthworms- Are known pulse of the soil. They play major role in maintenance of soil fertility and quality production. Earthworm is the chief operator of nature's microbe's factory. Local earthworms and uncountable microorganisms in the soil are the chief source of plants nutrition. They eat gravels, limestone organic biomass, move upwards and downwards making capillaries in the soil. This encourages better infiltration of water and helpful in upward movement of nutrients in available forms. Earthworms' excrement contains 5 times more of N, 7 times more of P, 11 times more K and 2 times more Mg and Ca than normal soil. Nitrogenous products of earthworm's metabolism are returned to the soil through casts, urine, micro proteins and dead worm's tissues. Other sources of nitrogen to the soils include decay of roots, excreta from invertebrate and vertebrate fauna, dead plant and animal material, and nitrogen fixation by microbes. Worm casts can promote lush growth of plants probably may be due to the plant growth promoters like cytokinin's and auxins present in the casts. Earthworms materially alter soil aeration, drainage, water absorption: and under certain circumstances limit soil erosion by reducing runoff. Earthworms play important role in minimizing root diseases. In conducive situations earthworm can add sizeable amount of fertile (cocoon) every year. These ultimately enhances activities of working soil organisms proliferates in their proximity. The most striking effect of earthworms may be the stimulation of microbial activity including *Actinomycetes, Streptomyces, Azotobacter, and Azospirilium,* thereby helpful in managing soil born fungal diseases and mediating N from biosphere. Earthworms release into soil certain vitamins and similar substances which may be the B group vitamin or free amino acids. Vermi wash promote lush growth of plants which may be due to plant growth promoters like cytokinin's and auxins present in the casts.

Cow- Cow has been regarded as *Kamdhenu* and most respected family member in ancient culture. The indigenous breeds of cow with hump at back, horns, dwells in neck are special features.

Food which permeates through rumen gets enriched with microbial consortia. Similarly, milk gets enriched with microbes residing in the cow udder and teats. All products of cow are of immense use in agriculture and human health. Need base bio enhancers, a new concept can be prepared by the cosmic farmer for enhancing soil fertility, crop productivity and quality. In cosmic farming, as external input only cow is needed. Two cows can help 4-5 hectares of cosmic area.

It is interesting to record that indigenous breed of cow (*Bos indicus*) with hump and horns have capacity of mediation of cosmic energy. Hump at the back of cow is of Pyramid shape, and horn act as antennae while belly is little replica of cosmos are special features of Indian cow.

Cow belly is mini replica of cosmos, horns have capacity to mediate solar energy. Cow rumen, udder and teats are rich in beneficial *Bacillus subtills* group microbes. Hence all products of cow i.e., dung, urine and milk are of much use in agriculture and human health. These cow products are used in preparation of bio enhancers, enriched composts and biopesticides.

Cow Aura Energy

Aura Energy is almost a new concept in cosmic farming, but there is immense possibility of use of cow- based products. Indigenous breeds with hump have maximum "Aura energy' light emitting from animate objects. Cow products- based formulations are used in various forms to address nutrient requirement, insect pest management. In fact, nutrient deficiency, pest and disease incidence occurs in plants, when Aura energy of soil and plants are depleted. Such plants show negative Aura energy which can be quantified with Universal Scanner.

Table-1 Aura energy of cow products and their role in agriculture

Cow product	Aura energy (m)	Impact
Milk	12-13m'	Owing to presence of A-2 is known as Amrita. It has unique blending of 101substances, 19 amino acids. Microbes like *Lacto bacillus* produces organic acids which promote crop growth and suppress pathogens. Has been used for seed treatment and management of viral infection during ancient time in India.
Cow curd	6.5-7m'	An important component in Panchagavya and good for human health. Fermented butter milk is used for seed treatment and for foliar application for the management viral diseases.
Cow ghee	14m'	Richest source of energy, helps in quick combustion of dung, by burning it cleanses air pollution. It is a power full vehicle for capturing solar energies and its transmission in larger area through Agnihotra fumes.

Cow product	Aura energy (m)	Impact
Cow dung	6m′	Richest source of beneficial microbial consortia. Contains Menthol, Ammonia, Phenol, Indole, Formalin. Effective soil conditioner, patties are used in Agnihotra and other Yajna. Has unique capacity of protection from nuclear radio activity and protection from bacterial infection. Rich source of beneficial microbes, used for seed coating and preparation of tree paste, composts and bio enhancers. Rich source of beneficial microbes, used for seed coating and preparation of tree paste, composts and bio enhancers.
Cow urine	8-9 m′	It is disinfectant, prophylactic, antibacterial, antifungal, antiviral, contains amino acids, cytokinin's, which play role in immunity enhancement and soil fertility. It is richest secretion from any source. It is disinfectant, prophylactic, antibacterial, antifungal, antiviral, contains amino acids, cytokinin's, which play role in immunity enhancement and soil fertility. Used in preparation of all bioenhancers.

Looking the current crises in agriculture, there is need for introspection of prevalent agricultural production system, reverting back to ancient agricultural practices based on mediation of cosmic energy through cow- based formulations.

Strange- Cow eats cocktail of food, but all the products i.e., the milk, dung and urine are free from any residues of toxic materials. It is due to our ignorance after stopping of milk, cows are left to face pathetic life on roads, towns and cities. One should remember that dung and urine are given by cow till her end, which can be used in preparation of specific bio enhancers and used in cosmic production of crops. Thus, it requires change of mindset of farmers and policy makers to convey strongly to take care of cow till end of her life and use their products in cosmic production of crops. Your small contribution will help country to get same name and fame as in ancient times.

Strange-*Cow lives on eating leftover material from human consumption say bhusha, paddy straw, sugar cane leaves. Grasses, oil cakes, chuni-chokar, but these are converted into valuable materials for agriculture and human health. This conversion is not possible by any factory or laboratory.*

Bio- enhancers: Bio-enhancers are organic preparations, obtained by active fermentation of animal & plant residues over specific duration. Bio enhancers are rich source of microbial consortia, macro and micronutrients and plant growth promoting substances. These are utilized to treat seeds/ seedlings, enrich the soil and induce better plant vigour.

Bio enhancers have been found quite effective in enhancing productivity of different crops and suppressing growth of various plant pathogens by producing anti-bacterial and anti-fungal compounds, hormones and siderophores. After working with organic farming systems over two decades in different capacity, I am with firm conviction that integration of Biodynamic and Homa organic farming which are supposed to be twin sisters, one giving emphasis for enhancement of rhizosphere and the other to the biosphere can address the current crises in farming. This I am trying to promote as "Cosmic Farming".

Table 2: Crop activities, cows-based formulation and their impacts

Synod.	Crop activity	Suitable bio-enhancer
i	Seed/plant parts treatment	Beejamrit /CPP (cow pat pit)/cow urine + dung powder + Agnihotra ash etc.
ii	Enhancing decomposition of biomass	Jeevamrit/CPP/Panchagavya/ Agnihotra ash water, Kunapjal, etc.
iii	Enhancing nutritive value of compost	CPP/Agnihotra ash enriched water, Biosol/ Vermi wash/ Kunapjal, Panchagavya etc.
iv	Enhancing soil fertility	Jeevamrit/ CPP /Biosol /Kunapjal Panchagavya etc.
v	Improvement in crop vigour	Panchagavya/ Kunapjal/ Biosol/ Vermi wash etc.
vi	Enhancing biotic & abiotic stress	CPP/Biosol/Kunapjal, Panchagavya, Vermi wash etc.
vii	Insect pest management	Kunapjal/Vermi wash/Biosol/ Dasparni Panchagavya, Biodynamic liquid pesticides.
viii	Seed/grain storage	Agnihotra ash/Panchagavya

Perusal of table-2 is an indication that for all farm activities cows- based formulations are available. Systematic research, keeping same breed and same feed of cows, their microbial profiling and impact in field in different agroclimatic situations are needed.

Out of number of Cows based organic farming systems being practiced by different farmers group, I propose two systems i.e., Biodynamic & Homa Organic Farming are twin sisters acting in complementary and supplementary mode. Major emphasis of biodynamic is to enhance biological activities of soil and Homa is effective in combating pollution of basic elements, therefore, integration of few select techniques of both in name of cosmic production is need of hour.

Biodynamic Farming

Bio Dynamics is the oldest organized alternative of agricultural movement in the world. The term *biodynamic* is taken from the two Greek words *bios* meaning life and *dynamics* meaning energy. Hence biodynamic farming refers to "working with the energies which create and maintain life." The basic principle of biodynamic farming is that when soil is balanced, then healthy plant will grow and transmit both substances and energy(life giving and life maintaining) as food. The aim of biodynamic concept is to establish a system that brings balance into all factors, which maintain life.

Rudolf Steiner (1861-1925) from Austria, theinventor of this unique technology gave the world, ideas on renewal of agriculture. Biodynamics mainly works on two poles: The earthy and the cosmic. The cosmic activities can work strongly through a living soil and so on into the plant. He emphasized that one must nourish the earth in such a way that the cosmic influences could continue to flow in. Using biodynamic preparations regularly opens the soil, to the to mediate cosmic energy and thereby, revitalize to work through the soil into the plants. The more biologically active the soil, the more beneficent forces can work through into plant and animals. It involves certain principles and practices for healthy soil, healthy plant and healthy food for human beings and feed for animals. In the system, energies from *cosmos, mother earth, cow* and *plants* are systematically and synergistically mediated. It is based on the knowledge that soil, plants, animals and men work together in one close agricultural cycle.

Biodynamic techniques, enhance, rejuvenate, to maintain soil quality. Cow horns and cow dung, after being buried together in the earth, make the most wonderful humus to spread on the land. This improves the structure of the soil dramatically and quickly.

Few special preparations are made from fermented manure (BD-500), quartz silica (BD-501) minerals, and select herbs (BD-502 to BD-507) over specific duration. These are used to help, restore and harmonize the vital life forces of the farm and to enhance the nutrition, quality and power of the food being produced. Steiner's basic philosophy is that plants grow not only through the fertility of soil but also with support from cosmos – the rhythms of the sun, moon, planets and wider constellations of the zodiac. Interaction between component of substance and energy factors form a balanced system are the basis for sustainable agriculture.

First group i.e., BD-500 and BD-501 both are prepared in cow horns. Cow horn is an ideal focal device to mediate the earthly and cosmic streams of force upon the materials packed in it.

Cow Horn Manure-BD-500 is basically fermented cow dung and is the basis for soil fertility and renewal of degraded soils. BD-500 is sprayed to enhance bio life in the soil by increasing humus content, increase soil porosity, enhance root development. It is usually the first preparation used during the changeover to organic/biodynamic system. Application of BD-500 vitalizes the soil and enhances the biological, physical and chemical soil properties.

Cow horn silica (BD-501)-Fine ground mountain quartz crystal (silica) fine paste is stuffed in cow horn shells and buried in ground during ascending sun and moon in March-April and taken out in September –October in ascending phase of moon. It is also known as fungal vaccine. It enhances the photosynthesis by higher chlorophyll content in foliage, strengthen cell wall to build strong plant structure and increases plant nutrients like K,P, and Ca uptake. Improves taste, colour, lustre, aroma and shelf life of produce.

Cow Pat Pit (CPP)

It is a special biodynamic field preparation is a strong soil conditioner. It is a multipurpose Polyculture. It is strong soil conditioner, enhances seed germination, promotes rooting in cuttings, improves in soil texture, provides resistance power to the plants against pests and diseases, replenishes and rectifies the trace element deficiency. It is used to improve soil fertility, plant vigour and foliar sprays, in preparation of tree paste and many other ways.

The other group of preparationsare (BD-502-BD-507) used in preparation of cow pat pit(CPP), and BD-compost, BD-liquid pesticides.These preparations bring in and enhance the, cosmic forces from planets and also catalyses few specific macro and micro nutrients thereby, enhancing efficacy of system better than other organic farming systems. These preparations are associated with specific planets such as Venus, Mercury, Mars, Moon, Jupiter and Saturn. These are used to enhance the mediation capacity of cosmic energy in compost, Cow Pat Pit, BD-compost and BD-pesticides.

Homa Therapy

Currently to humanity is facing environmental pollution. Most of international efforts have not been its dividend and as a result climatic disaster in the one or other parts of world is routine. In order to combat environmental pollution, Shree Gajanan Swami (1918- 1987) who is known as **Kalki Avatar** (to combat environmental pollution), after intense efforts conceived simple techniques of Homa Therapy, capable of fighting with pollution with number of other associated benefits. It consists of use of specific size of pyramid made from copper, cow dung patties, cow ghee, unbroken rice and chanting two mantras at sun set and sunrise of the place. Agnihotra is the main pillar in Homa Therapy. Agni means- fire and Horta means -purification. Historical perspective of Agnihotra. In Valmiki Ramayana, there is reference that Bhagavan Shree Ram was daily performing Agnihotra in his life. On the day of crown ceremony, both Mata Sita and Ram performed Agnihotra.

Agnihotra is a gift to humanity from ancient- most Vedic Sciences of bio energy, medicine, agriculture and climate engineering. It is based on principle that "you heal the atmosphere and healed atmosphere will heal everyone". Agnihotra is science of Pyramid logy, biorhythm of nature, burning of organic substances and sonic power of chanting specific mantras. When Agnihotra is performed, the Agnihotra smoke gathers practices of harmful radiation from the biosphere.

Materials Required in Agnihotra

Materials required for Agnihotra can be arranged by anyone, anywhere in the world with little efforts. As on today, it has spread in more than 100 countries in all continents, mainly by amateur practitioners. It is not associated with religion, sex, education or caste and race. It is matter of great concern that regular performance of Agnihotra is possible by every family in the world.

Brief Account of Agnihotra

Pyramid: Pyramid is made from copper is universally acknowledged for its excellent conduction of electricity and heat. The word 'pyramid' means fire in the middle. Pyramid has special power to attract cosmic & solar energy. This ancient –word meaning for pyramid is closely connected with the inexplicable energies emanating from the central shape. The pyramid shape is widely experienced to generate and store a special energy field, which possess bacteriostatic properties.

The inverted pyramid shape *allows* controlled generation and multidimensional dissipation of energy. It acts as a generator of unusual energy fields and spreads in its surrounding areas wherever Resonance Point (RP) has been established. The Agnihotra has an obvious link with fire, heat, electro-magnetic force and because of its inverted pyramid shape have connection with the cosmic energy fields. The dimensions of the pyramid are: 14.50 cm x 14.50 cm at the top, 5.25 x 5.25 cm at the bottom and 6.25 cm in height with three steps.

Rice

Rice is widely grown in all part of the world. Brown i.e., unpolished, unbroken rice (*Akshata*) is used for performance of Agnihotra. Highly polished rice loses nutritional value and hence brown rice is used. If rice is broken, the subtle energy structure is disturbed and it is not fit for Agnihotra healing fire. If organically produced rice is available, it is to be preferred.

Dried Cow Dung Cakes

Cow dung is treated as medicine in all ancient culture whether they may be Indians of North or South Americans, Scandinavians, East or West Europeans, Africans and Asians. Cow's dung contains Menthol, Phenol, Formalin, Phosphoric acid, Potash, Ammonia, Indole, and Nitrogen. Make a pan cake like patties and dry them in sun. Agnihotra fire is to be prepared from this dried cow dung.

Cow Ghee

Russian scientist Sorovich claimed that burning of 10 gram of cow ghee generates one tone of oxygen. Cow ghee comprises of glycerol and number of saturated and unsaturated fatty acids such as Steric acid, Oleic acid, Linolenic acid, Palmitic acid, Myristic acid and Laric acid. When ghee is burnt along with other materials in prescribed disciplines, it becomes much more energetic. Traditionally, homemade ghee has been found ideal for performance of Agnihotra. This can last without refrigeration for long time. Ghee is a very special medicinal substance. It helps in rapid combustion of the materials and keeps the fire alight. On combustion and oxidation, these form hydrocarbons, aldehydes and formaldehyde responsible for fragrance and purification of the area by minimizing pathogenic bacteria.

Agnihotra Timings

"At sunrise the many fires, electricity's, ethers and more subtle energies emanating from the sun extend all the way to the earth and produce a flood effect at those coordinates where the sun is said to rise. It is awesome. The flood enlivens and purifies everything in its path, destroying what is impure in its wake. This torrent of life sustaining energies causes all life to rejoice. At sunrise that music can be heard. The morning Agnihotra Manta is essence of that music. It is quintessential sound of that flood. At sunset the flood recedes".

The time component of Agnihotra is one of the most essential aspects to be observed. Light radiated by sun at sunrise and sunset is termed as diffused light in the sky which has great ecological significance. Since sun at the rise or set becomes near to horizon in transverse/ oblique position, solar rays have to cover a longer distance to reach to the earth and in this course solal rays (electromagnetic waves) of visible spectrum of lower wave lengths i.e., violet, indigo, and blue and probably ultraviolet, X-rays and Y-rays are scattered and lost in the atmosphere. The other visible rays of higher wavelengths i.e., yellow, orange, red and infra-red of invisible spectrum reach to the earth and the sky looks yellowish red at the time of sunrise and sunset. One can say that solar ray reaching to the earth at both times contains greater proportion of infra-red which is lethal but more penetrating and greater energetic in nature.

Mantras

There are vibrations that exist everywhere. It is only vibrations when someone tries to understand deep into it. Where there is vibration there is also sound. When someone performs Mantras, the sounds activate special vibrations that will create specific atmosphere for its impact. Then the desired results are realized. These vibrations exist for everything, so anything can be activated, controlled or changed by Mantras. When one utters the Mantra with a pure mind into the pyramid at Agnihotra time, the ash gets energized from cosmic energy retains that energy and healing properties of the ash become more powerful.

Agnihotra Mantras

At Sunrise

Soory'aya Sw'ah'a, Soory'aya Idam Na Mama

(Add the first portion of rice mixed with ghee into the fire)

Praj'apataye Sw'ah'a, *Praj'apataye Idam Na Mama*

(Add the second portion of rice mixed with ghee into the fire)

This completes morning Agnihotra.

At Sunset

Agnaya Sw'ah'a, Agnaye Idam Na Mama

(Add the first portion of rice mixed with ghee into the fire)

Praj'apataye Sw'ah'a, Praja'pataye Idam Na Mama

This completes evening Agnihotra

(a' is pronounced like a in 'father')

For timings and other information following websites will be helpful

www.homatherapyindia.com; www.agnihotra.com.au; www.cosmicfarmingindia.com.

One should try to observe exact times for performance of Agnihotra. After Agnihotra, try to spare as many minutes as one can for meditation. One should sit at least till the fire extinguish. It creates pollution free healing atmosphere.

Resonance Point

With establishment of Resonance point, one can manage up to 80 hectares area, as 'Cosmic Village' can be managed with almost same human efforts. After performance of morning Agnihotra, its energy is such powerful, touches the four corner and get boot up to 12 km in the biosphere. The resultant Agnihotra ash is full of subtle energy. It is useful in agriculture, cattle and also for human health.

Agnihotra ash is full of subtle energy with lot of therapeutic values to plants, animals and human. It is used in powder form or enriched water and a potent bio enhancer, known as Biosol. Agnihotra ash is used for seed storage, treatment of seed/seedling, enhance soil fertility, water quality, availability, and manage pest and disease in plants.

Use of Agnihotra Ash

- Mix Agnihotra ash with grains and seeds before their storage. It keeps the attack of pests at zero level. The grain can be used either for human consumption or as seed.
- One kg Agnihotra ash is spread over one acre land before crop sowing/ planting. The ash not only helps to maintain soil fertility but also energies the soil to mediate cosmic energies that aid in plant growth.
- Agnihotra ash increases the moisture retaining capacity of the soil.
- It renders the nutrients and elements already existing in the soluble and easily available to plant in great extent.
- Agnihotra ash has been observed to be rich source of microbes i.e., P- solubilizers and N-fixtures hence are effective to release soluble phosphates from the soil and make them bio available.
- Equally effective to enhance availability of N from atmosphere,
- It plays important role in the growth of plants.
- For seed treatment place seeds fully covered in Cow urine for 20-30 minutes, drench urine and carefully cover it with dung powder and Agnihotra ash before sowing. Seeds with hard seed coat can be kept for 1-2 hours.
- Five grams of Agnihotra ash can place in hills before transplanting vegetable/ flowers and herbal seedlings.
- In fruit plant, incorporate 25-50 g Agnihotra ash before planting.
- Fine powder of Agnihotra ash can be dusted on foliage, fruits wherever insect infection is problem.
- Agnihotra ash can is mixed with organic mulch all-round the tree trunk. It will help in quick decomposition of mulches and provide desired nutrients required by the plants.
- In perennial fruits mulching can be done with its own dropped foliage. It is advisable that some legumes are sown in the orchard where some solar

radiation is available, mix with dropped foliage and make a thick mulch leaving tree trunk and drench it with Agnihotra ash water/Biosol.

- An excellent plant food can be made by making a solution of Agnihotra ash+ cow urine and water. For this, incorporate one Kg Agnihotra ash + 5 liters cow urine in 200 liters water. Incubate the mixture for three days and stir it thrice a day. After filtration Agnihotra enriched water can be used as foliar spray or drenching on mulches.
- One liter of this solution can be sprayed over a compost heap of 5X1X 1M. The heap should be covered with gunny bag/ sugarcane trash/ coconut fronts or paddy straw.
- This will be helpful in quick decomposition of non-decomposed material and improve nutritive value of compost.
- Agnihotra ash is placed in a gunny bag and kept at main water source used for irrigation of r field or orchard.
- One kg Agnihotra ash is placed at weekly interval in wells/ponds used for irrigation of field. This will be helpful in improving in water quality and also availability.
- A thick tree paste can be prepared by incorporating cow dung+ cow urine+ Agnihotra ash and clay.
- In fruits like mango, cashew, litchi, guava, aonla, sapota etc. the paste should be smeared on the trunk, if possible, on main branches and cut ends.
- Twice pasting i.e., once immediately after fruit harvest and the other in spring –February-March will help in management of borers, gummosis and will help in better vascular activities.
- Use of Agnihotra ash in Vermi composting- In process earthworms should be put in the compost material at the end of the afternoon and moistened with Agnihotra ash solution after sunset.

Biosol

A mixture of cow dung, Vermi compost, cow urine and Agnihotra ash mixed in water is placed and properly sealed. At weekly interval upper valve is opened to release the gas. In this way in four weeks Biosol is ready.

Biosol has been observed to be rich source of beneficial bacteria, fungi, actinomycetes, N-fixers, P solubilizers. Impact of Biosol has been reported in cabbage and tomato. Enhanced availability of macro and micro nutrients, and soil biological attributes. Biosol has been found effective in enhancing quality production and management of pests and diseases.

It is pertinent to mention that there are certain commonalities, between these systems, the common practices are discussed briefly.

These are rich source of beneficial microbial consortia as evident from-table-3.

Microbial profiling of BD-products and Homa products i.e., Biosol at CISH, Lucknow indicated interesting information as given in table-3.

Microorganisms (cfuml1)	BD-500	BD-501	CPP	Biosol
Bacteria	3.75xX102	10.5x	1.75x102	31.39x 102
Fungi	6.98x105	11.33x105	8.33x105	30.10x102
Actinomycetes	15.5x106	3.33X106	13.13x106	33.1x105
Pseudomonas	1.3x106	4.9x105	4.9x105	105.6x105
Rhizobium	3.89x105	24.25x105	8.40x106	12.25x105
P-SM	22.48x105	4.9x105	23.37x 106	2.96x104
Azotobacter	nil	Nil	224x106	2.96x104
Az spirillum	6.0x107	4.8x107	31.00x106	Nil

Perusal of table-3 is inconformity that all these preparations are potent source of beneficial microbial consortia, play important role in farming. Two bioenhancers i.e., Cow Pat Pit from Biodynamic farming and Biosol from Homa organic farming are potent bio enhancer of immense value. These can be prepared by every farm family, by keeping few humped cows at the farm.

Biodynamic farming- BD-500, BD-501, Cow Pat Pit, BD-Compost, BD-Pesticides.

Homa Organic Farming- Agnihotra ash, Agnihotra ash enriched water and Biosol. It is relevant to mention that specific bioenhancers are available for all crop activities as enumerated in table-4.

Table 4: Crop activities and perspective use of bio-enhancers

Sunoo.	Crop activity	Suitable bio-enhancer
i	Seed/plant parts treatment	Beejamrit /CPP (cow pat pit)/cow urine + dung powder + Agnihotra ash etc.
ii	Enhancing decomposition of biomass	Jeevamrit/CPP/Panchagavya/ Agnihotra ash water, Kunapjal, etc.
iii	Enhancing nutritive value of compost	CPP/Agnihotra ash enriched water, Biosol/ Vermi wash/ Kunapjal, Panchagavya etc.
iv	Enhancing soil fertility	Jeevamrit/CPP/Biosol/Kunapjal/ Panchagavya etc.
v	Improvement in crop vigour	Panchagavya/Kunapjal/Biosol/ Vermi wash etc.
vi	Enhancing biotic & abiotic stress	CPP/Biosol/Kunapjal, Panchagavya, Vermi wash etc.

Sunoo.	Crop activity	Suitable bio-enhancer
vii	Insect pest management	Kunapjal/ Vermi wash/Biosol/ Dasparni Panchagavya, Biodynamic liquid pesticides.
viii	Seed/grain storage	Agnihotra ash/Panchagavya

Perusal of table-4 is indicative that specific bioenhancers are available for all crop activities. Few meager financial supportsin terms of infrastructure and skill upgradation, particularly farm women will have unimaginable implications.

There is intense interplay in microbial world, if saprophytic dominate, will manage nutrition and take care of health of plants from pest and disease infection, while pathogenic one, willresult weak plants, invite pest and disease infection. Therefore, all efforts need to be concentrated to enhance activities of saprophytic microbial consortia. Agnihotra ash is rich in subtle energy, its incorporation enhances N-fixtures and P-solubilizers many folds. Efforts should be made to sow/plant crops on rows & furrow system, facing north-south direction to trap effective electromagnetic and solar radiation, minimizing water requirement with improved production.

Conclusion

Cosmic Farming is the integration of two organic systems i.e., Biodynamic Farming & Homa Organic Farming. Both are twin sisters acting in complementary and supplementary mode. Major emphasis of biodynamics is concentrated to enhance soil biology, while homa is effective in combating pollution of natural elements and boosting all production factors i.e., soil, availability of nutrients, water quality & availability, biodiversity of fauna, flora, particularly earthworms' activity in soil and bees in biosphere, thereby quality production in all crops and situations.

Cosmic Farming is particularly suited to small and marginal farmers of India, situated in for of places, facing resource constraints. Similarly, the current situation of humped dry cows is pathetic in the country. I am of strong view that Government of India and number of state government in recent decades are promoting organic farming with number of subsidized projects. Looking huge subsidy for its promotion, like other government supported project number of private players have crept in manufacturing of number of organic formulations. Once, the project is withdrawn, everyone forgets. This is major concern for introspection by every stake holder.*It reminds the Chinese saying that 'if someone is interested to support any group, he shouldn't give fish, but equip and train them for fishing*.Therefore, I proposethat,a part of finances being incurred as subsidy for fertilizers, should be diverted to 'equip, train and tune' people for assertive promotion of cosmic farming.

Cosmic Farming can ensure high quality production, nutritional security, mitigating the soil; water and environmental crises are the need of the hour. Hence its assertive promotion will have immense value. With corona pandemic, assertive promotion of cosmic farming and local consumption of cosmic food can resolve many crises.This will have for reaching implications in providing effective technique for future generation.

Foods for Pranic Body: Living food with a lot of Prana or Vital energy present in pure water, clean air, organic foods etc. Food for Astral/ Emotional Body: Positive emotions, caresses, tenderness, kindness respect, joys, smile, laughter, certain physical gesture, etc. Nature will be constantly benefiting from the Homa Atmosphere makes its signals even more tangible at specific movements.

It is praiseworthy that, currently country is giving emphasis for utilization of solar energy, instead of fossil fuel or imported source of energy. As result of emphasis by the government, country is moving fast, to meet the energy need of the country. In recent Cop-26 conference, PM has emphasized for solar alliance to be world leader. Now few corporate sectors are putting major resources for establishment of solar panel to capture solar energy. It is very expensive venture and their life expectancy will be limited, disposal of these solar panels may pose again problem of their disposal. While in promotion of 'Cosmic Farming in Horticultural Crops' will provide unique opportunity to country like India, being dominated by small and marginal farmers, in remote corner of country constrained by resources.

Foods for Pranic Body: Living food with a lot of Prana or Vital energy present in pure water, clean air, organic foods etc. Food for Astral/ Emotional Body: Positive emotions, caresses, tenderness, kindness respect, joys, smile, laughter, certain physical gesture, etc. Nature will be constantly benefiting from the Homa Atmosphere makes its signals even more tangible at specific movements.

Lesson to learn-One should prefer to reside in cosmic environment rich with biodiversity, eat cosmically raised vegan food, lives as per cosmic rhythm's way for healthy, happy and harmonious life. Horticultural crops are well suited for cosmic production. Current time i.e., 'Amrit Kaal' is challenging period particularly for agriculture fraternity to get them involve in intensive promotion of cosmic farming, in their capacity. This will be in consonance of Prime Ministers initiatives for Sustainable Development Goal (SDG) and -LiFEStyle for Ecofriendly Environment. This will ultimately result cherishing the dreams of: One Earth: One Family: One Future thereby recognition of India 'Vishwa Guru'.

References

Proctor Peter (2008). Biodynamic Farming and Gardening, The Classic Text on Biodynamic Agriculture, The Organic Farming Sourcebook, OFAI, Mapusa, Bardez, Goa-403507.

Pathak, R.K. and N. Dube (in press) Cosmic Farming in Horticultural Crops, Pub. Westville Publishing House, New Delhi,

Paranjpe V.(2023). Homa Therapy Our Last Chance, Published by: Fivefold Path Publications, Shreeniwas, 40, Ashok Nagar, Dhule-424001, Maharashtra, India, Email: fivefoldpathpub@gmail.com

Source for training and materials for Biodynamic Farming, Supa Biotech Pvt.Ltd., Dina Lodge Compound, MallitalNainital Cantonment, Uttarakhand

Source for Homa materials, Shree Abhay Paranjpe, 40, Ashok Nagar, Dhule, 424001, Maharashtra, India, Email fivefoldpath@gmail.com

Shree Sarvajit Paranjpe, Fivefold Path Mission, Homa Therapy Goshala, Ladvi, Mandleshwar-451221, Maheshwar, Dist. Khargone,Madhya Pradesh.

https://www.youtube.com/siddhagirimath Panch Tatva Sumangalam

https://youtu.be/BdjA8qRTh1s)-Ram Biodynamic Farming

https://www.youtube.com/watch?v=hICGX5uK0Dg&t=1041s- BD- Farming Hortak-49, Pathak and Shah

http://homatherapy.org/agnihotra-a-talk-by-shree-vasant-paranjpe-part-1/- Homa Farming

Chapter 11

On-Farm Production of Quality Organic Inputs for Entrepreneurship Development in Horticultural Crops

R.A. Ram*, Govind Kumar, Israr Ahmad, Shailendra Kumar Maurya[4] and A. Kumar[5]
Central Institute for Subtropical Horticulture, Rehmankhera, Lucknow-226101
**Ex- Principal Scientist & Head, Crop Production (Act.)*
**E-mail: raram_cish@yahoo.co.in*

Use of agrochemicals was initiated in commercial cultivation of horticultural crops after advent of green revolution. Significant increase in production, productivity and area expansion under horticulture crops has occurred during the past two decades under the influence of chemical intensive systems (Pathak and Ram, 2004). Imbalanced use of chemical fertilizers, especially nitrogenous has resulted in some regions manifesting adverse effects on the environment, polluting soil and ground water resources. Soil quality, especially that of organic matter and micro-nutrients deficiencies are becoming ubiquitous, threatening sustainability and quality of produce impacting nutritional security. Further, indiscriminate use of pesticides, has led to development of resistance to pests to pesticides, while destroying irretrievably the beneficial ones viz., honey-bees, pollinators, parasitoids, predators, besides causing harmful pesticide residues in the end product adversely impacting productivity and food safety. Increasing consciousness about conservation of environment as well as of health hazards caused by agrochemicals has brought a major shift in consumer preference towards food, quality, particularly in the developed countries. Global consumers are increasingly looking forward to organic food that is considered safe and hazard-free. As per the excellent taste, wide variability with respect to varieties, nutritive, therapeutic values, and organic production of horticultural crops will have immense scope in domestic and export markets. Preliminary attempts in organic production in different crops in various regions have shown spectacular response. It is pertinent to record that with proper follow up all the organic production systems i.e. Biodynamic Farming, *Rishi Krishi,* Nateuco *Krishi, Panchagavya Krishi,* Homa organic farming and Natural Farming have indicated that high quality production without the use of agrochemicals. There is need to integrate few compactable techniques from these established system to develop package of practices which can easily be adopted by the common growers to assure sustainable organic production. Availability of quality organic inputs in the markets are not assured and sometimes they do not meet the standard set by the BIS. Emphasis should be given for on farm production quality organic inputs for cost effective and sustainable production. There are few organic farming practices in which all the inputs viz; composts, bio-enhancers and bio-pesticides are produced with locally available materials. Composts are rich source of nutrients, bio-enhancers are rich source are of beneficial microbes and bio-pesticides are effective source for the management of various insect pests.

On Farm Production of Organic Inputs

Compost

Importance of Organic Matter

Crop waste burning is one of the common practices in today's farming system and in northern India. These organic wastes can easily be converted in to rich composts like biodynamic, vermi, NADEP and vermi compost etc. to improve the soil fertility. Biodynamic compost is an effective soil conditioner and is an immediate source of nutrients for crops. It can be prepared by using green (nitrogenous material) and dry leaves (carbonaceous material) in 40-60 ratio. Integration of cow dung slurry and BD-502-507 in the compost enhances the decomposition process. The composition of air, moisture and warmth is very important in the breakdown and decomposition of the organic materials. The enrich compost gets ready in 3-4 months depending upon the prevailing temperature and moisture. Vermi compost can easily be produced with use of cattle dung and locally available bio mass. If vermi beds are provided a little slope say 5-10 %, and then vermi wash can also be obtained without any expenditure. In an experiment, application of biodynamic compost (30 kg/tree) + bio-enhancers (CPP 100 g, BD–500 and BD 501 as soil and foliar spray) in 30 years old tree of mango cv. Mallika yielded 160 kg fruit per tree with improvement in fruit quality (TSS (26.36 ^{0}Brix), total caretonoids (6.40mg/100g) and fluorescence recovery after photo bleaching (FRAP) (74.478 micromole/litre) with BC ratio of 5.10 (Ram *et al*, 2019). For NADEP composting, a 2x3.30x1.25 m brick aerobic structure is constructed at elevated place in farm area and organic wastes, cow dung and fertile soil are placed in the structure layer wise. After filling of organic wastes 1.5 feet above the structure, heap is pasted with cowdung and mud. Due to incorporation of fertile soil, proper ratio of green and dry biomass aerobic composting process, nutritional status of end product is better than the farm yard manure. Farm wastes (cow dung, green/dry grasses, wheat/paddy straw and weeds) along with good quality farm soils are used in composting. Microbe mediated compost is prepared with microbial consortium which contains lactic acid and photosynthetic bacteria, yeast, filamentous fungi and actinomycetes etc. These are aerobic and anaerobic in nature and can exist in acidic and saline soil both. These effective microbes are used for production of good quality compost with dung and farm wastes.

Numbers of composting methods have been standardized in various parts of the country. Any of this can be attempted as per facilities available at the farm. These composts contain more nutrients than common farmyard manure (Fig.1).

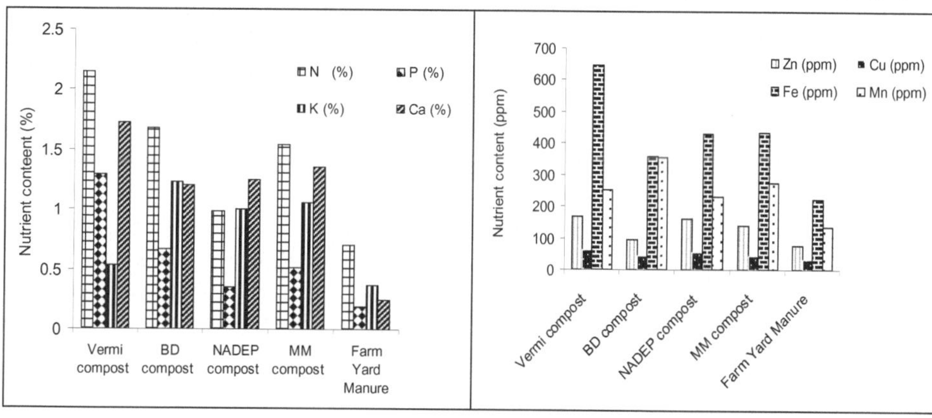

Fig.1: Major and micro nutrients level in various compost

Bio-enhancers

Vrikshayurveda (science of plant life) deals with growing and nourishing plants with liquid manures (prepared with plant and animal products). There are several verses (e.g. Brahatasamhita of Varahamira, Surpala's Vrikshayurveda and Upavanavinoda of Sharangadhara), which provide information on irrigation methods using water mixed with herbal products obtained from different plant species and animal products to increase crop production (Sadhale, 1996). Kunapajala is a combination of two words 'Kunapa' which means dead or decaying matter and 'Jala' means water derived from this dead matter according to Sanskrit dictionaries. Kunapajala generally promotes flowering, fruiting and vegetative growth and is also used for plant protection measures (Majumdar, 1935). Preparation of *Kunapajala* which involves boiling of flesh, fat and marrow of animals such as deer, pig, fish, sheep, goat in water, placing it in earthen pot, and adding milk, powders of sesame oil cake, black gram boiled in honey, decoction of pulses, ghee and hot water used to be the common booster of plant vigour (Nene, 2007). *Kunapajala* is sprayed on plant to enhance its vigour and production. Since research explanation and supporting data on implication of *Kunapajala* are not available to support their use in present scenario. In a study Sarkar, *et al*, 2014 have reported that Panchagavya and Kunapajala individually as well as in combination proved their efficacy in promoting the growth and yield of vegetables crops.

Concentrated manures, bio products in powder or in liquid form, henceforth termed as Bio-enhancers are organic preparations, obtained by active fermentation of animal and plant residues over specific duration. These are rich source of microbial consortia, macro, micronutrients and plant growth promoting substances including immunity enhancers. These are utilized to treat seeds/seedlings, enhance decomposition of organic materials thereby enrich soil and induce better plant vigour. These could be a potent tool to utilize these in fertigation in various crops (Pathak and Ram, 2012).

1. Cow Dung

The use of cow dung has been indicated since the time of Kautilya (C.300 BC) and it was used for dressing seeds, plastering cut ends of vegetative propagated sugar cane, dressing of wounds, sprinkling of diluted solution on crops since ancient times. ICAR-Central Institute for Subtropical Horticulture, Lucknow, India has identified presence of 4 potent strains of *Bacillus subtilis*, which have shown strong anti pathogencity against number of diseases in fruits like mango, guava, papaya rots (Pathak *et al*, 2009). They have also identified actinomycetes as *Streptosporangium pseudovulgare* which has shown anti pathogenic potential against *Colletotrichum gloeosporioides* (anthracnose pathogen) and *L. theobromae* (gummosis, stem end rot and die back pathogens. Microbial analysis of fresh cow dung showed that it contains nitrogen fixing, phosphorus solubilizing bacteria and others beneficial microbes in plenty (Ram *et al*, 2019).

2. Cow Urine

The use of cow urine is known for a long time in India. *Gaw-mutra* (cow's urine) has been described as a liquid with innumerable therapeutic values, capable of curing several incurable diseases in human beings and plants. Cow urine is used for preparation of number of bio enhancers and bio-pesticides, which are effective in improving soil fertility, quick decomposition of organic wastes and management of large number of pests and diseases in varied group of crops.

3. Cow Horn Manure (BD-500)

It is basically fermented cow dung and is the basis for soil fertility and renewal of degraded soils. Fresh cow dung filled in cow horn is buried in humus rich soil by integrating cosmic energies for specific period. These are buried in September - October and taken out in March - April when cosmic forces are more active underground. Microbial analysis of BD-500 showed that it contains beneficial microbes responsible for improving soil fertility.

4. BD-501

BD-501 is prepared by incubating silica powder in cow horn during April – October in ascending period of Moon. For maximum effect, the BD-501 should be applied once at the beginning of a plant's life, at the four-leaf stage and again at the flowering stage or fruit maturation stage. BD-501 should be applied on the leaves in the form of fine 'mist' during morning at sunrise and on the best constellation is Moon is opposite to Saturn. Spraying of BD-501 has significantly improved the fruit quality of mango (Pandey, *et al*, 2003). Microbes isolated from cow horn silica (BD-501) (Ram and Pathak, 2017) are presented in Fig.2.

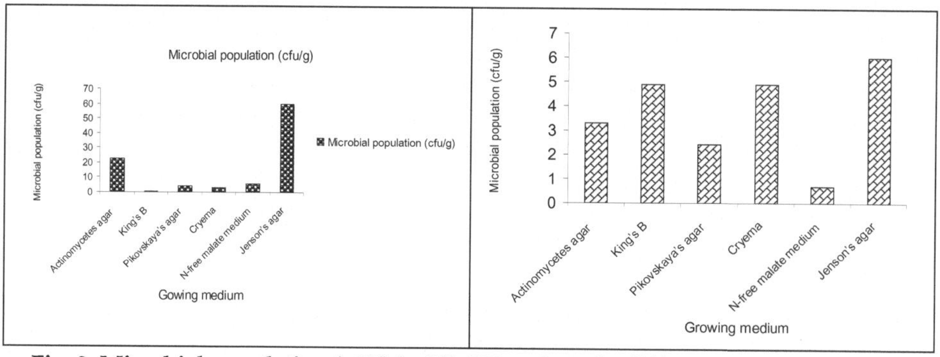

Fig. 2: Microbial population (x10[6]) in BD-500 and 501 in different growing medium

5. Cow Pat Pit (CPP)

It is a special biodynamic field preparation and also called as 'soil shampoo'. Fresh cow dung collected from lactating and pasture going cows is used for preparation. Cow dung is fermented along with crushed egg shells (calcium) and basalt/bentonite (clay) dust duly mixed and placed in pit size of 3 x 2 x 1.5'. Two sets of BD-502-507 preparations are incorporated for catalyzing the composting process. It is a concentrated source of beneficial organisms. In a study, CPP showed highest bacterial load (4.8 X 10^6) per g, *Rhizobium* (1.9 X $10^{6)}$, *Azospirillum* (0.2 X $10^{6)}$, *Azotobacter* (8.0 X 10^5) and fungi (2.5 X 10^6) (Ram *et al.*, 2010). It also contained the highest amount of *B. subtilis* (1.9 X 10^6) responsible for disease tolerance (Proctor, 2008). Microbes isolated from the cow pat pit are presented in Fig.3.

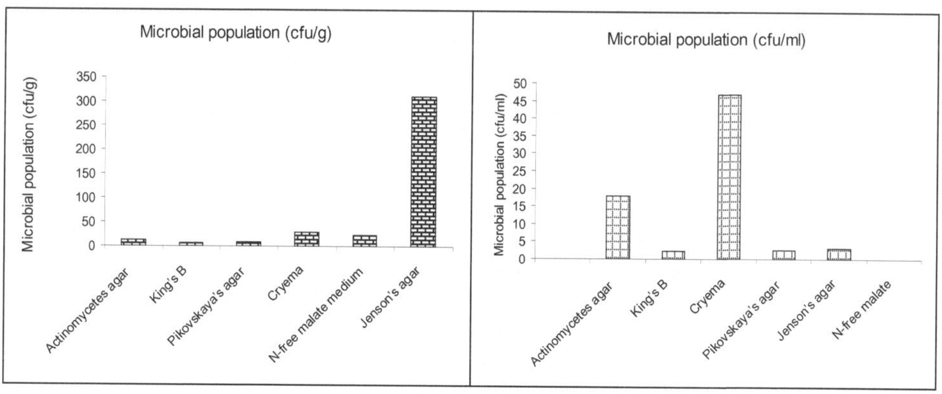

Fig.3. Microbial population (x10[6]) in cow pat pit and Panchagavya on various medium

6. Biodynamic Liquid Pesticides

Biodynamic liquid pesticides are prepared by materials i.e. cow dung, urine and leaves of leguminous tree, neem leaves, fish waste, caster leaves and other medicinal plant parts. Besides cow dung, cow urine and one set of BD preparations (502-507) are also incorporated. The liquid manures are used to promote the vigour and quality production. On an average, preparation of liquid manure takes 8-10 days. In an experiment, mango hopper management with biodynamic liquid pesticides was effectively done. Before spray, the hopper population was 3.07 hoppers panicle^{-1} and after the spray the reduction in the hopper population was found up to 15th SMW (Standard Meteorological Week) with 0.95 hoppers panicle^{-1}. Second spray was taken up at 14th SMW, as result hopper population reduced to 0.4 hoppers panicle^{-1} up to 19th SMW (Ram et al, 2019).

Panchagavya

This preparation is rich in nutrients, auxins, gibberellins and microbial fauna and acts as tonic to enrich the soil to induce plant vigour with quality production. It is prepared with five production cow (cow dung, cow urine, cow milk, cow ghee, cow milk curd) along with banana, coconut ware and palm wine. Its gets ready in 18 days. Panchagavya is equally effective for all types of plants, milch animals, goat, poultry, fish and pet animals. Its remarkable effects have been demonstrated in fruits like mango, guava, acid lime, banana, spice turmeric, flower-jasmine and vegetables such as cucumber spinach etc. The spray of panchagavya on chilies produces dark green coloured leaves within 10 days and its role has been reported by Sreenivasa *et al*; 2009. The effective micro organisms in panchagavya were the mixed culture of naturally occurring, beneficial microbes, mostly lactic acid bacteria (*Lactobacillus*), yeast (*Saccharomyces*), actinomycetes (*Streptomyces*), photosynthetic bacteria (*Rhodopsuedomonas*) and certain fungi (*Aspergillus*). In view of the fact that panchagavya contains naturally occurring beneficial microorganisms (Swaminathan, 2005), some of which are nitrogen fixers and P-solubilizers (Sreenivas, *et al*, 2011), can be considered as an ideal organic growth promoter. However it is advisable to use within 30 days of its preparation to achieve better effect (Patnaik *et al*, 2012). Application of panchagavya has been found more profitable than recommended doe of fertilizer and agrochemicals spray. Kumar *et al*, (2015) reported that among the different treatments tested against the management of insect pest in teak (*Tectona grandis*), 7% and 5% diluted panchagavya application were found to be more effective in controlling the pests. Cost effective analysis showed 7% diluted Panchagavya was found to be cheaper and affordable to the tree growers.

Systematic microbial analysis of Panchagavya from 0 to 25 days suggests that Panchagavya contains maximum microbial load on 18th day of preparation. Therefore, use of Panchagavya on 18th days will be more effective than other days (Ram *et al*, 2017) (Fig.3).

7. Jeevamrita

Jeevamrita is prepared by fermenting 10 kg cow dung, 5 liters urine, 2 kg jaggery, 2 kg pulse flour, 100 g virgin soil in 200 liters of water by simple facilities created at the farm with minimum expenditure. Credit for development of recipes for Jeevamrita and its extensive use goes to Palekar (2006). Its can be used at 15 to 30 days interval through irrigation water coupled with mulching (green/dry {monocot + di-cot}) and proper soil aeration. Jeevamrita is a rich bio-formulation contains consortia of beneficial microbes (Ram *et al*, 2018). Systematic microbial analysis of Jeevamrita from the 0 day to 20th days of preparation suggests that formulation should be used within 6-9 days for maximum benefits (Fig.4) Ram *et al*, 2017. It can be drenched on mulch either by drip irrigation or through spraying. It is also effective in quick decomposition of crop residues if applied with irrigation water given for field preparation. With micro irrigation, 3 to 4 times more area can be covered with 200 liter of Jeevamrita.

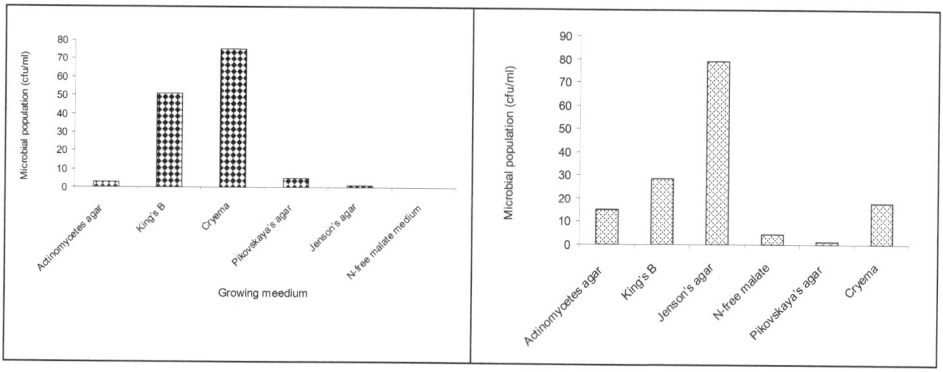

Fig.4. Microbial population ($x10^6$) in Jeevamrita and Beejamrita in various medium

8. Beejamrita

It is a special bio-enhancer prepared with locally available materials for seed and seedlings treatment. As preparation is very cheap and cost effective, can easily be prepared and used by small and marginal farmers. It prepared using 5 kg cowdung, 5 liters cow urine, 50g vigin soil, 50 g slacked lime and 20 liters of water. It is advisable to prefer earthen/mud pot of suitable size for the preparation. It is advisable to prepare fresh *Beejamrita* and use it for treatment of seeds/seedlings and other plant parts before sowing/planting/transplanting. Microbial analysis of Beejamrita showed that it should be used on 7th days of preparation. In case the crop is transplanted it is advisable that seeds are treated with *Beejamrita* before sowing and while planting the root portion of the seedlings are dipped before transplanting. Microbial analysis of Beejamrita is presented in fig.4.

9. Amritpani

This bio-formulation is rich in nutrients and beneficial microbes. Ingredients for preparation of amritpani and its intensive use were advocated by Deshpandey (2003). It is used to improve seed germination, soil fertility and plant vigour. Systematic microbial analysis of Amritpani from 0-20 days suggests that Amritpani should be utilized from 6-9th days of preparation for maximum effectiveness. After application in soil it improves humus content, earthworm activity and thus soil fertility and crop productivity. Microbial analysis of Amritpani is presented in Fig.5.

Table 1: Materials required for preparation of Amritpani

S.No	Ingredients	Quantity
1	Cow dung	10kg
2	Cow ghee	250g
3	Honey	500 g
4	Water	200 liters

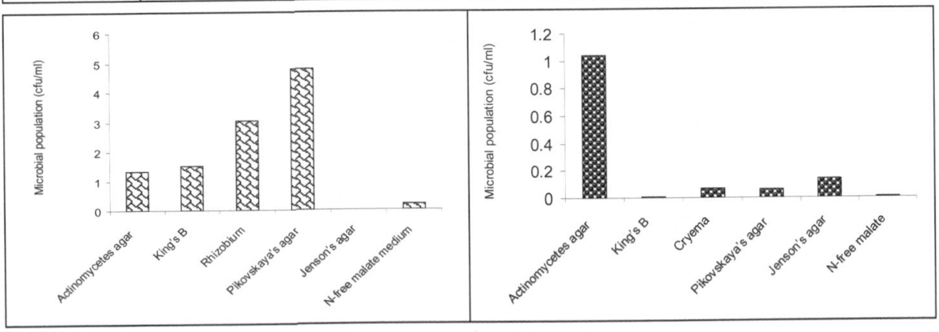

Fig. 5: Microbial population (x10^6) in Amritpani and Vermiwash in various medium

9. Vermi Wash

Vermi wash is a liquid leachate obtained by excess water to saturate the vermi composting substrate. It is collection of excretory products and mucus recreations of earthworms along with nutrients from the soil organic molecules. Vermiwash was found to contain enzyme cocktail of proteases, amylases, urease and phosphatase. Microbiological study of vermiwash revealed that it contains nitrogen-fixing bacteria like *Azotobacte*r sp., *Agrobacterium* sp. and *Rhizobium* sp. and some phosphate solublizing bacteria. Laboratory scale trial showed effectiveness of vermiwash on cowpea plant growth. (Zambare *et al*, 2008).Vermi wash contained total heterotrophs i.e.; *Nitrosomonas* 10.1 x 10^3, *Nitrobacter* 1.12 X 10^3 and total fungi, 1.46 X 10^3.

Recently microbial study of vermi wash revealed that it contains nitrogen-fixing bacteria like *Azotobacter* sp., *Agrobacterium sp., Rhizobium* sp., and phosphate solubilzing bacteria. Protease in soils helps in seed germination, while amylases help in availability of simple carbon source for enhancement of plant vigour and productivity as well. Soil born micro flora is essential for growth of plants because organic nitrogenous compounds and phosphorus are decomposed and mineralized by fixing and phosphate solubilzing bacteria (Fig.5). Presence of large number of beneficial microorganisms helps in plant growth and protects from a number of pathogen in the field. Repeated spray of vermi wash has been found effective even in management of thrips and mites in chilies (George *et al*, 2007). If needed, vermi wash may be mixed with cow urine (1:1:8 ratio, vermi wash, cow urine and water), and used as foliar spray for nutrients and pesticidal properties.

References

Deshpande, M. D. (2003). Organic Farming wrt. Cosmic energy, Non Violence Rishi – Krishi. Khede-Ajra, Kholapur, Maharashtra, p 65.

Pandey, G.; B. P. Singh; Ram, R. A. and Pathak, R. K. (2003). Studies on pre-harvest spray of BD 501 on fruit quality and shelf life of mango (*Mangifera indica* L.). National Symposium on Organic Farming in Horticulture for Sustainable Production, held at CISH, Lucknow, 29 -30 August, 2003, p 74.

Garg, N.; Om Prakash and Pathak, R.K. (2003). Cow Dung: A source of potent Bio control agents. National Seminar on Cow in Agriculture and human health. Agri History Society, Udaipur, Rajasthan, p 82

Majumdar, Girija. (1935). Upavana Vinodha. The Indian Research Institute, Calcutta, p 58.

Nene, Y.L. (2007). Utilizing traditional knowledge in agriculture, National Seminar on Organic Agriculture: Hope of Posterity, July, 13-14, 2007 organized by UPCAR and NCOF, pp 6-10.

Palekar, S. (2006). The Philosophy of Spiritual Farming, (Zero budgets Natural Farming, Amrit Subash Palekar, Amravati, Maharashtra.

Pathak R K and Ram R A. 2004. *Manual on Vedic Krishi*, pp-1-38.

Pathak, R. K. and Ram, R. A. (2009). Nutrients levels of different compost. Manual on Jaivik Krishi. Bulletin No.37. pp.31.

Pathak, R.K. and Ram, R.A. (2012). Bio enhancer: A Potential tool to Enhance Soil Fertility and Crop Productivity, pp 84-86. Lead talk in Souvenir & Abstracts in International Conference on Organic Farming for Sustainable Horti-Agriculture and Trade Fair, organized by Jharkhand State Horticulture Mission, 8- 9 November, 2012.

Patnaik, H. P.; Dash, S. K. and Shailaja, B. (2012). Microbial composition of Panchagavya, Journal of Eco-friendly Agriculture 7 (2): 101-103.

Proctor, P. 2008. Biodynamic Farming and Gardening, Other India Press, Goa, India.

Ram, R. A. and Pathak, R. K. (2017). Bio-enhancers, LAP Lambert Academy Publishing, Germany, ISBN 978-3-330-33128-0

R. A. Ram; Verma Anil; Gundappa and V. Supriya. (2017). Studies on yield, fruit quality and economics of organic production of mango cv. Mallika. Innovative Research on organic Agriculture, WOC, New Delhi (ISBN: 978-3-86576-177-4).

Pathak, R.K. and Ram, R.A. (2013). Bio-enhancers: A potential tool to improve soil fertility and plant health in organic production of horticultural crops, Progressive Horticulture, 45(2):237-254.

Ram, R. A. and Verma, A. K. (2017). Energy input, output and economic analysis in organic production of guava (*Psidium guajava*) cv. Allahabad Safeda. *Indian Journal of Agricultural Sciences*, **87** (4): 462-67.

Ram, R. A.; Singha A. and Singh, Kumar, Vinod. (2019). Improvement in yield and fruit quality of mango (*Mangifera indica*) with organic amendments *Indian Journal of Agricultural Sciences,* **89** (9):1429-33.

Ram, R. A.; Singha, A. and Vaish, S. (2018). Microbial characterization of on-farm produced bio-enhancers used in organic farming. *Ind. J. Agric. Sci.* **88** (1): *Indian Journal of Agricultural Sciences,* **88** (1): 35-40.

Sadhale Nalini. (1996). Surpalas Vrikshayurved. Asian Agri-History Foundation, Secunderabad (A.P.): 35-39.

Sarkar, S.; Kundu, S.S. and Ghorai, D. (2014). Validation of ancient liquid organics-Panchagavya and Kunapajala as plant growth promoters. Indian Journal of Traditional Knowledge, **13**(2): 398-403.

Sreenivasa, M.N.; Nagaraj, N. and Bhat, S.N. (2009), Beneficial traits of microbial isolates of organic liquid manures – First Asian PGPR Congress for Sustainable Agriculture, 21-24 June, 2009, ANGARU, Hyderabad.

Sreenivasa, M.N.; Naik, N. and Bhat, S. N. (2011). Nutrient status and microbial load of different organic liquid manures. Karnataka Journal of Agricultural Sciences, **24**: 583-584.

Swaminathan, C. (2005). Food Production through Vrikshayurveda way. In *Technology for Natural Farming Eds.* Agricultural College & Research Institute, Madurai, Tamilnadu, India. 18-22 pp.

Zambare, V.P.; Padul, M. V. Yadav, A. A. and Shete, T.B. (2008). Vermi wash: Biochemical and Microbial Approach as Eco friendly soil conditioner, *ARPN J. Agric & Biol. Sci*; 3(4): 1-4.

Chapter 12

Natural and Integrated Bio-natural Farming Technologies for Horticulture

S. K. Shukla*, P. K. Shukla, Govind Kumar, Tarun Adak, T. Damodaran and Alok Shukla
ICAR-Central Institute for Subtropical Horticulture, Rehmankhera, Kakori, Lucknow - 226101
**Email: skshuklacish@gmail.com*

Safe production of fruits and vegetables is a challenge due to indiscriminate use of chemical fertilizers and pesticides. It is very common to see the use of even banned pesticides for managing insect pests. Efforts have been made during last two decades to promote organic farming which is a system of farm design and management to create an ecosystem of agriculture production without the use of synthetic external inputs such as chemical fertilisers, pesticides and synthetic hormones or genetically modified organisms. Some of the organic farming systems have been developed from time to time such as Biodynamic Farming, Rishi Krishi, Panchgavya Krishi, Natural Ecological (NATUECO) Farming and Homa Farming, etc (Ram and Pathak, 2016).

Organic farming is gaining popularity throughout the world. Research shows that there are around 30 percent more wildlife and plants near ecological production fields as compared to conventional farming. This is because there are no pesticides, and fertilizer is used far less. According to FiBL survey 2021 India holds a unique position among 187 countries practicing organic agriculture. India is home to 30% of total organic producers in the world having 2.30 million ha. Total organic cultivation area, 27, 59,660 total farmers (11, 60,650 PGS and 15, 99,010 India Organic), 1703 total processors and 745 traders.

The problems being faced by farmers in adoption of organic farming are also important. Organic farming is claimed to be more expensive as it requires using large amounts of organic manures/fertilizers/external inputs on the land brought from those who sell them. It also requires activities like mixing of manures/compost, ploughing and tilling and has an effect on the surrounding environment. Organic farming involves use of permitted chemicals and composting of organic waste and farm yard manure (FYM) for application to the soil for soil nutrition. For last few years, Natural Farming is being promoted by Govt of India as an alternative to organic farming.

What is Natural Farming?

Natural farming is defined as a "Chemical- free livestock based farming". Soundly grounded in agro-ecology, it is a diversified farming system that integrates crops, trees and livestock, allowing the optimum use of functional biodiversity. It holds the promise of enhancing farmers' income while delivering many other benefits, such as restoration of soil fertility and environmental health, and mitigating and/or reducing greenhouse gas emissions. This farming approach was introduced by Masanobu Fukuoka, a Japanese farmer and philosopher, in his 1975 book "The One-Straw Revolution". Internationally, Natural Farming is considered a form of regenerative agriculture which is a prominent strategy to save the planet. In India, Natural farming is promoted as Bhartiya Prakritik Krishi PaddhatiProgramme (BPKP) under Paramparagat Krishi Vikas Yojana (PKVY). BPKP is aimed at promoting traditional indigenous practices which reduce externally purchased inputs.

Natural farming, also known as "the Fukuoka Method", or "do-nothing farming", is an ecological farming approach established by Masanobu Fukuoka (Fukuoka, 1978). Fukuoka was a Japanese farmer and philosopher who introduced the term "Natural Farming" in his book *"The One-Straw Revolution"* published in 1975. The concept referred not to lack of effort, but to the avoidance of manufactured inputs and equipment. Natural farming is related to fertility farming, organic farming, sustainable agriculture, agro-ecology, agroforestry, eco-agriculture and perma-culture, but should be distinguished from biodynamic agriculture.

Popularly known as "Do nothing" farming, natural farming is an environmentally regenerative way of growing food, founded not in a technique, but in an equitable relationship between farmer and nature. It is this relationship that forms the foundation of the natural farmer's actions.

Natural farming contrasts starkly with most contemporary forms of farming in that it is premised on developing closer relationships between farmer, land, and consumer, where as the industrial farming process—including industrial scale organic—forces a fundamental separation of these relationships. While techniques and methods vary widely, natural farmers share common roots in re-connecting themselves and their farming processes to the Earth, in learning from nature directly, and in cultivating food that regenerates the health of the natural world and of humanity.

The Original Concept of Natural Farming

The system works along with the natural biodiversity of each farmed area, encouraging the complexity of living organisms, both plant and animal that shape each particular ecosystem to thrive along with food plants. Fukuoka saw farming both as a means of producing food and as an aesthetic or spiritual approach to life, the ultimate goal of which was, "the cultivation and perfection of human beings".

He suggested that farmers could benefit from closely observing local conditions. Natural farming is a closed system, one that demands no human-supplied inputs and mimics nature.

Fukuoka's natural farming practice rejected the use of modern technology, and after twenty-five years, his farm demonstrated consistently comparable yields to that of the most technologically advanced farms in Japan, doing so without the pollution, soil loss, energy consumption, and environmental degradation inherent in these modern types of farming. One of the main prompts of natural farming, is to ask why we should apply modern technology to the process of growing food, if nature is capable of achieving similar yields without the negative side-effects of these technologies. Such ideas radically challenged conventions that are core to modern agro-industries; instead of promoting import of nutrients and chemicals, he suggested an approach that takes advantage of the local environment. Although natural farming is sometimes considered a subset of organic farming, it differs greatly from conventional organic farming, which Fukuoka considered to be another modern technique that disturbs nature.

Fukuoka claimed that his approach prevents water pollution, biodiversity loss and soil erosion, while providing ample amounts of food, and there is a growing body of scientific work in fields like agro-ecology and regenerative agriculture, that lend support to these claims.

Masanobu Fukuoka's Principles of Natural Farming

In principle, practitioners of natural farming maintain that it is not a technique but a view, or a way of seeing ourselves as a part of nature, rather than separate from or above it. Accordingly, the methods themselves vary widely depending on culture and local conditions.

Rather than offering a structured method, Fukuoka distilled the natural farming mindset into five principles:

i. No tillage

ii. No fertilizer

iii. No pesticides or herbicides

iv. No weeding

v. No pruning

Principally, natural farming minimizes human labour and adopts, as closely as practical, nature's production of foods such as rice, barley, citrus in biodiverse agricultural ecosystems. Without ploughing, seeds germinate well on the surface if site conditions meet the needs of the seeds placed there. Fukuoka used the presence of spiders in his fields as a key performance indicator of sustainability. Fukuoka specifies that the ground remain covered by weeds, white clover, alfalfa, herbaceous legumes, and sometimes deliberately sown herbaceous plants. Ground cover is present along with grain, vegetable crops and orchards. Chickens run free in orchards and ducks and carp populate rice fields.

Periodically ground layer plants including weeds may be cut and left on the surface, returning their nutrients to the soil, while suppressing weed growth. This also facilitates the sowing of seeds in the same area because the dense ground layer hides the seeds from animals such as birds.

For summer rice and winter barley grain crops, ground cover enhances nitrogen fixation. Straw from the previous crop mulches the topsoil. Each grain crop is sown before the previous one is harvested by broadcasting the seed among the standing crop. Later, this method was reduced to a single direct seeding of clover, barley and rice over the standing heads of rice. The result is a denser crop of smaller, but highly productive and stronger plants.

Fukuoka's practice and philosophy emphasized small scale operation and challenged the need for mechanised farming techniques for high productivity, efficiency and economies of scale. While his family's farm was larger than the Japanese average, he used one field of grain crops as a small-scale example of his system.

Yoshikazu Kawaguchi at Akame Natural Farm School

Widely regarded as the leading practitioner of the second-generation of natural farmers, Yoshikazu Kawaguchi is the instigator of Akame Natural Farm School, and a related network of volunteer-based "no-tuition" natural farming schools in Japan that numbers 40 locations and more than 900 concurrent students. Although Kawaguchi's practice is based on Fukuoka's principles, his methods differ notably from those of Fukuoka. He re-states the core values of natural farming as:

- Do not plow the fields
- Weeds and insects are not your enemies
- There is no need to add fertilizers
- Adjust the foods you grow based on your local climate and conditions

Kawaguchi's recognition outside of Japan has become wider after his appearance as the central character in the documentary *Final Straw: Food, Earth, Happiness*, through which his interviews were translated into several languages. Since 2016, Kawaguchi is no longer directly instructing at the Akame school which he founded. He is still actively teaching however, holding open farm days at his own natural farm in Nara prefecture.

No Till

Natural farming recognizes soils as a fundamental natural asset. Ancient soils possess physical and chemical attributes that render them capable of generating and supporting life abundance. It can be argued that tilling actually degrades the delicate balance of a climax soil:

- **Tilling may destroy crucial physical characteristics of a soil** such as water suction, its ability to send moisture upwards, even during dry spells. The effect is due to pressure differences between soil areas. Furthermore, tilling most certainly destroys soil horizons and hence disrupts the established flow of nutrients. A study suggests that reduced tillage preserves the crop residues on the top of the soil, allowing organic matter to be formed more easily and hence increasing the total organic carbon and nitrogen when compared to conventional tillage. The increases in organic carbon and nitrogen increase aerobic, facultative anaerobic and anaerobic bacteria populations.

- **Tilling over-pumps oxygen to local soil residents**, such as bacteria and fungi. As a result, the chemistry of the soil changes. Biological decomposition accelerates and the microbiota mass increases at the expense of other organic matter, adversely affecting most plants, including trees and vegetables. For plants to thrive a certain quantity of organic matter (around 5%) must be present in the soil.
- **Tilling uproots all the plants in the area, turning their roots into food for bacteria and fungi**. This damages their ability to aerate the soil. Living roots drill millions of tiny holes in the soil and thus provide oxygen. They also create room for beneficial insects and annelids (the phylum of worms). Some types of roots contribute directly to soil fertility by funding a mutualistic relationship with certain kinds of bacteria (most famously the Rhizobium) that can fix nitrogen.

Fukuoka advocated avoiding any change in the natural landscape. Fukuoka avoided the creation of terraces in his farm, even though terraces were common in China and Japan in his time. Instead, he prevented soil erosion by simply growing trees and shrubs on slopes. Ladybirds consume aphids and are considered beneficial by natural farmers that apply biological control.

Some Other Forms of Natural Farming in India

Rishi Kheti

In India, natural farming of Masanobu Fukuoka was called "Rishi Kheti" by practitioners like Partap Aggarwal. The Rishi Kheti use cow products like buttermilk, milk, curd and its waste urine for preparing growth promoters. The Rishi Kheti is considered to be non-violent farming without any usage of chemical fertilizer and pesticides. They obtain high quality natural or organic produce having medicinal values. Today still a small number of farmers in Madhya Pradesh, Punjab, Maharashtra and Andhra Pradesh, Tamil Nadu use this farming method in India.

Zero Budget Farming

Zero Budget Farming is a variation on natural farming developed in, and primarily practiced in southern India. It is also called spiritual farming. The method involves mulching, intercropping, and the use of several preparations like jeevamrit, beejamrit, neemastra, agnyastra, dashparni ark, brahmastra, etc. which include cow dung and/or cow urine. These preparations, generated on-site, are central to the practice, and said to promote microbe and earthworm activity in the soil. Indian agriculturist Subhash Palekar has researched and written extensively on this method (Devvrat, 2017, Devvratand Palekar, 2019; Palekar, 2023).

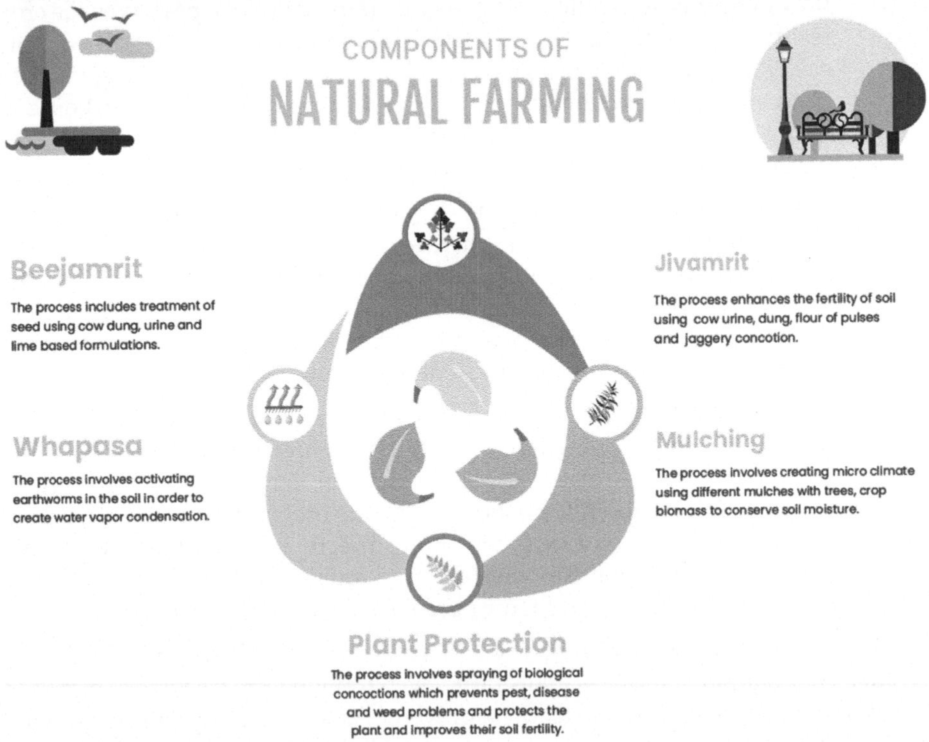

Habitat Management is key to Success of Natural Farming (Pathak and Ram, 2019)

- Role of plants in capturing cosmic & solar energy (micro & macro flora) is enormous.
- Both micro & macro flora captures cosmic & solar energy.
- CO_2 in wood & roots is fixed permanently (Ecology).
- Plants have only organic diet, i.e. in oxidized ,chelate and ionic form which is accomplished by microbes.
- Plants restrict soil erosion, aerate soil & dead rootlets provide organic matter to soil organisms.
- Primary source of biomass.
- Soil beneath the canopy of few trees is rich in beneficial microbial consortia.
- Few plant parts have pesticide properties utilized for pest management
- Farming System, multi crops, rotation, mixed crops.
- Green manures, cover crops, inter crops, pump crops and mulching have many implications.
- Trees such as casuarinas, tamarind, neem, madhuca help in maintenance of soil biology and environmental ecology.

- Growing fodder and fuel plants on field bunds and in vacant area.
- Species grown in association should have different growth habits and different needs for light.
- Auro greens which can add large amount of organic matter quickly.
- Plants require 30+ nutrients for satisfactory performance.
- Plants get 98 to 98.5 % nutrients from, air, water and solar energy and need hardly 1.5-2% from soil.
- Atmosphere is being polluted by most human activities.
- The NPK, Fe, S and Ca are available in the soil but not in available form. They need to be transformed through action of microbes (Bacteria, microbes and local earthworms) already present in soil but excessive use of chemicals has destroyed them.
- In zero budget farming (ZBF) nothing has to be purchased from outside.

Importance of Natural Farming

- **Minimized Cost of Production:** It is **considered as a cost- effective farming practice** with scope for raising employment and rural development.
- **Ensures Better Health:** As Natural Farming does not use any synthetic chemicals, **health risks and hazards are eliminated.** The food has higher nutrition density and therefore offers better health benefits.
- **Employment Generation:** It generates employment on account of natural farming input enterprises, value addition, marketing in local areas, etc. The surplus from natural farming is invested in the village itself. As it has the potential to generate employment, thereby stemming the migration of rural youth.
- **Environment Conservation:** It ensures better soil biology, improved agrobiodiversity and a more judicious usage of water with much smaller carbon and nitrogen footprints.
- **Livestock Sustainability:** The integration of livestock in the farming system plays an important role in Natural farming and helps in restoring the ecosystem. Eco Friendly bio-inputs, such as Jivamrit and Beejamrit, are prepared from cow dung and urine, and other natural products.
- **Resilience:** The changes in soil structure with the help of organic carbon, no/low tillage and plant diversity are supporting plant growth even under extreme situations like severe **droughts and withstanding severe flood** and wind damage during **cyclones.** NF impacts many farmers positively by imparting resilience to the crops against weather extremities.

M. Fukuoka in Japan started Natural Farming and obtained crop yields comparable with chemical farming without any soil erosion. Basic principle is to reduce use of external inputs to the farm land which degenerates soil nature. He prepared natural insecticide from chrysanthemum roots called pyrethrum and used against cabbage worms and moths. Soil is supplemented with the microbial inoculums like *Beejamrit, Jeevamrit* to accelerate the propagation of soil microflora, helpful to soil enrichment.

Practices

- No tillage or minimum tillage
- Use of organic mulching
- Use of Jeevamrit and GhanJeevamrit
- Use of Beejamrit for seed/seedling treatment
- Maintenance of proper balance of air and water in soil's pore spaces (WAFSA)
- N fixing and PSB help in improving the soil fertility and plant growth.

Advantages

- Physical work and labour are minimized
- Yields may be less initially but comparable over the period with chemical farming
- Increase in soil fertility over the years
- Water requirement reduces significantly
- **Has no side effects:Is it realistic**? Yes, because it is based on regeneration of earth's habitats and resources. No system of production can be sustainable if it constantly takes more resources from the earth Industrial scale farming is absolutely, unquestionably unrealistic.

Organic and Natural Farming – A Comparison

Aspects	Organic Farming	Natural Farming
Cost	Organic farming is more expensive than NF because it requires using large amounts of organic manures/fertilizers on the land brought from those who sell them.	Cost of farming gets reduced as it encourages decomposition of organic matter through biological processes after mulching.
Effort	Requires activities like mixing of manures/compost, ploughing and tilling.	Decomposition of organic matter by microbes and earthworms is encouraged right on the soil surface itself, which gradually releases nutrients into the soil.
Environmental impact	Has an effect on the surrounding environment .	It conforms to local biodiversity.
Composting	OF wants permitted chemicals and composting of organic waste and FYM for application to the soil for soil nutrition.	NF doesn't want any chemicals or organic compost inputs like vermiculture. It promotes a natural catalyst of biological activity in the soil and natural protection form diseases.

Bio-natural farming : The concept of bio-natural farming includes the principles of natural farming coupled with canopy management and integrated disease pest management using non-chemical methods and use of bio-formulations.It includes use of polythene banding for mealy bugs, use of pheromone traps for fruit flies, green manuring, mulching, use of chemicals permissible in organic farming, centre opening of improving light penetration, etc

Experiences with Integrated Bio-natural Farming at ICAR-CISH, Lucknow with Regard to Mango Cultivation

The ICAR-CISH has just started research on natural farming at Raebareli Road Campus of the Institute. Different systems of Chemical/traditional, natural and integrated bio-natural farming are being compared. Initial soil samples were taken and soil organic carbon ranged from 0.4-0.6. All the concepts of natural farming like total absence or minimum interference by tillage, maintenance of organic mulch, promoting in-situ decomposition of organic matter instead of composting and use of various cow based organic products like *jeevamrit, ghanjeevamrit, amritpani, neemastra,* plant based liquid manures, *agnyastra and dashparni* ark. Under integrated bio-natural farming, efforts were made to open up the tree canopies through centre opening and also removing very low, cris-cross and the branches infested with *Loranthus*. Polythene banding and spray of *Beauveria bassiana* around tree trunk during first week of January were used to manage mealy bugs successfully. A 10-15 cm thick organic mulch in tree basins was ensured with organic waste available at the farm. *Jeevamrit* and *Amritpanti* were used for drenching organic mulch around the trees from time to time. Pollination and fruit set were significantly better when compared with chemical/traditional farming. The protective sprays (two with *Neemastra*@ 3%, *Agnyastra* @3% and *Dashparni Ark* @ 3% and two sprays with cow urine @ 10%) were done at 7 days interval during panicle emergence and development stages. The incidence of blossom blight and powdery mildew was negligible. However, population buildup of mango hopper and thrips was recorded very high during fruit set and early development of fruits. The crop was successfully protected from thrips and hoppers by one spray of Agrospray oil 97% EC (1%) followed by one spray with neem oil emulsion (0.3%) 7 days after the first spray. Later, one spray each with vermi-wash, cow urine and Gliricidia based liquid manure was also done. Significant increase in fruit set was observed at naturally maintained farm when compared with the traditionally maintained mango orchards managed with recommended fertilizers and pesticides. Heavy pollinator population (4-12 per panicle) was recorded during full bloom which improved fruit set and incidence of *Jhumka* (fruit clustering), the physiological disorder was minimum. Frequent presence of birds like peacock, heron and crow was observed in the natural farm when compared with rare presence in chemically maintained mango orchards in Malihabad mango belt. With the prevailing fruit set, a fruit yield of 8-14 MT/ha in Dashehari was obtained when compared with the yields of chemical farming (4-6 MT/ha). Fruit size in Dashehari (160-231 g/fruit) was smaller in case of natural farming which was probably due to excessive fruit set when compared with chemical farming (185 to 351 g/fruit). The TSS was in the range of 19-21°B, acidity from 0.213 to 0.234 % and ascorbic acid content was in between 31.25 to 39.58 mg/100g which showed non-significant differences when compared with chemical farming. The initial results of natural farming are extremely encouraging and will certainly go a long way in safe fruit production of mangoes in the years to come. Various workers have recorded positive effects of application of vermwashon soil and plant (Zambare et al, 2008; Venkataramana et al., 2009) on various crops. Sreenivasa et al. (2020) found beejamrit a potential source of beneficial bacteria.

Conclusions

Enrich the soil with organic input, nutrients and microbial population to make soil healthy and suitable for higher production. Intercropping, recycling of crop residue, mulching with locally available biomass, frequent use of GhanJeevamrit in place of FYM/Compost, use of Jeevamrit during field preparation & irrigation and 4-6 times foliar spray of Jeemamrit/ cow urine/Vermi wash,as per convenience, without using a gram of agrochemicals, crop productivity can be sustained in most fruit trees.

References

Devavrat (2017) Zero Budget Natural Farming (in Hindi). Pushpak Press Pvt Ltd, Okhla Industrial Area, New Delhi : pp56

Devavrat, A. And Palekar, S.. (2019) KamLagatPrakritikKrishi. Pushpak Press Pvt Ltd, NaiDelhi : pp166(https://atarikanpur.icar.gov.in/devrat.pdf)

Fukuoka, M (1978) [1975 Sep.] – *The One-Straw Revolution: An Introduction to Natural Farming*, translators Chris Pearce, Tsune Kurosawa and Larry Korn, Rodale Press.

Palekar, S. (2023). Zero Budget Prakritik (Adhyatmik) Kheti. (https://spnfhp.in/Upload/MediaGallery/PDF/3/2023-Jan-05-18-43-35-794.pdf)

Pathak, R.K. and R.A. Ram, (2019). A concept paper on Cosmic Nutrient Management in horticultural crops, *Progressive Hort.*, 51(2): 111-115.

Ram, R. A. and Pathak, R. K. (2016). Organic approaches for sustainable production of horticultural crops: A Review. *Progressive Horticulture*, 48(1):1-16

Sreenivasa, M.N.; Nagaraj, N.; and Bhat, S.N. (2010). Beejamrutha: A source for beneficial bacteria, *Karnataka J.Agril. Sci*, 17(4) 731-735.

Venkataramana, P.; Narasimhamurthy, Rao, B.; K and Kamble, C.K. (2009). Efficacy of foliar sprays of vermin-wash and cow dung wash on biochemical and yield attributes in mulberry (*Morus alba* L.) Karnataka *Journal of Agriculture Sciences*. 22:921-923.

Zambare, V.P.; Padul, M.V, Yadav, A. A; and Shete, T.B. (2008). Vermi wash: Biochemical and Microbial Approach as Eco friendly soil conditioner, *ARPN J. Agric& Biol. Sci*; 3(4): 1-4.

Chapter 13

Transforming Horticulture through Hydroponics Technologies

Ankur Agarwal*, Om Prakash, Basant Ballabh and Devkanta P. Singh

Defence Institute of Bio-Energy Research (DIBER), DRDO, Haldwani, Distt Nainital, P.O. Arjunpur, Uttarakhand- 263139, India

**Email: ankurdr@rediffmail.com*

The challenge to agriculture in the next few decades is to achieve maximum production of food without further irreversible depletion or destruction of our natural resources commensurating with increasing population. Nearly all of this population increase will occur in developing countries. Urbanization will continue at an accelerated pace, and about 70 percent of the world's population will be urban (compared to 49 percent today). According to FAO report, by 2050 the area of arable land will be expanded by 70 million hectares, or about 5 percent. Globally, irrigated agriculture covers one fifth of arable land and contributes nearly 50 percent of crop production although country wise distribution is very much skewed than the average (FAO, 2009). To address these problems, various innovative technologies are being promoted globally which allows judicious utilization of land and water resources. Soil-less/ hydroponics and vertical farming has been one such technology (Pradhan & Deo, 2019; NAAS, 2019).

According to a report published during 2016 based on the Census of 2011, out of 16793 villages of Uttarakhand, about 1053 villages have no inhabitants and another 405 have a population of less than 10. Out of these depopulated villages, 14 are within an aerial distance of 5 km from the international borders. This is a grim situation of migration from border villages adding to the worries of Indian Security agencies. Among the various factors reported for migration few notable are search for better livelihood and less productivity from agriculture due to changing weather and problem of wild animals. Hence, measures are required to increase the livelihood opportunities in these border areas to curb the problem of migration through intervention of modern technologies. Among these technologies, soil-less cultivation (hydroponics) have shown potential for profitable agriculture in varying environments from mountainous region to Antarctica as researched by

DIBER (DRDO) during past several decades (Pant *et. al.*, 2018; Agarwal *et. al.*, 2019a; Agarwal *et. al.*, 2019b). Hydroponics is a technique of growing plants in nutrient solution without using soil. It is a simple yet effective technology. The present day hydroponics systems are more customized for user requirement and are available of the shelf. Hydroponics system have been successfully customized by DIBER for various altitudes such as Haldwani (foot hill), Pithoragarh (5000′ asl) and Auli (9000′ asl). The technology has been successfully demonstrated during Antarctica expeditions during 90s.

Advantages

The hydroponic system has been used primarily for growing vegetables, cut flower saplings, potted foliage plants as well as grasses with several advantages such as

- A soil-less system with 70-80 percent less requirement of water
- Judicious utilization and optimization of nutrients for growing multiple crops in single nutrient solution,
- More production per unit area with additional possibility of vertical farming leading to space conservation
- Protection from pest or diseases being devoid of soil and under controlled condition
- Organic agriculture can be achieved
- Spaces available at roof tops and indoor can be productively utilized
- Fresh farm produce can be in the vicinity of user (reducing transportation cost)
- Cultivation possible in remote and harsh areas where soil cultivation is not possible

The key advantages, which hydroponics technology can offer in the livelihood opportunity development in the border villages of Uttarakhand are protection from wild animals. Most of the time due to topographical challenges it is not possible to codon the farm area with fencing and thus make it vulnerable to attack by stray and grazing animals vis-à-vis wild animals. Secondly, possibility of vertical farming allows farming in a compact area with substantial water saving due to reuse and recycling and hence dependency on rain can be minimized for farming. Third advantage of round the year production possibility make it more promising for these border areas where natural calamities and inclement weather keeps the area land locked and not condusive for open field agriculture.

Cost of Various Hydroponics Unit

Cost of hydroponics unit depend upon the local availability of materials, size of the unit to accommodate plants, requirement of automation, quality of water, etc. customization hydroponics unit generally being used in commercial farms are fully automatic

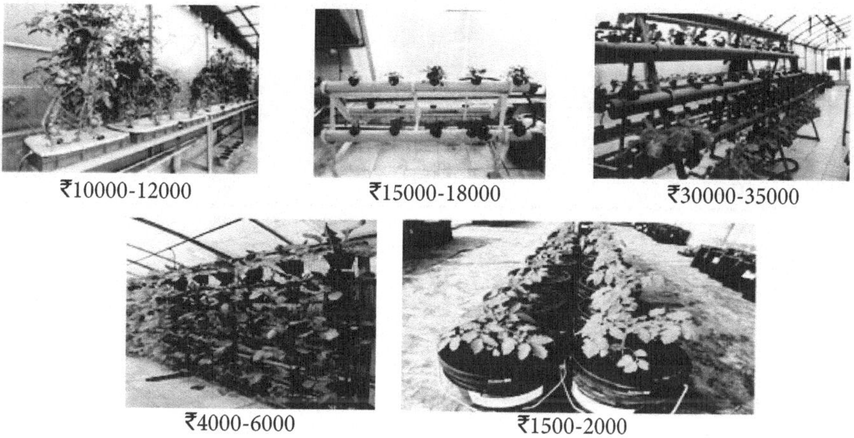

Fig. 1: Cost of various hydroponics unit for 100 sq. ft. area

Global Scenario

The global hydroponics market was valued at about 6.9 billion U.S. dollars in 2016, and is forecasted to grow to about 12 billion U.S. dollars by 2025. Expected annual growth 6.5% during the forecast period, 2018 to 2023. Europe is traditionally the largest market that is implementing advanced techniques in hydroponics. Asia-Pacific forms the second largest market for hydroponics. Globally, leafy greens have major share among the crops grown in hydroponics (Fig. 1).

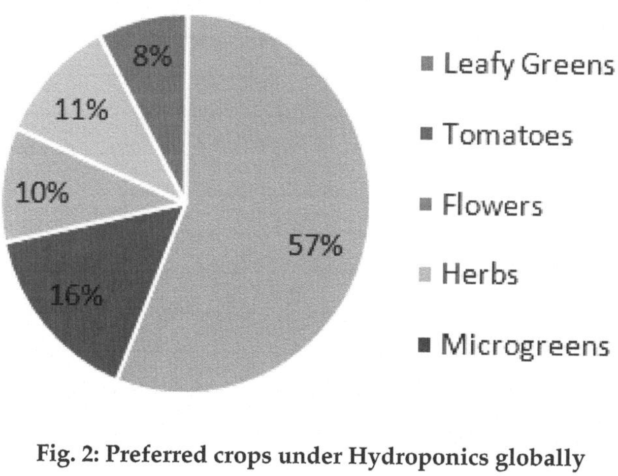

Fig. 2: Preferred crops under Hydroponics globally

Productivity of Various Crops under Hydroponics

The present paper highlights the efforts made by DIBER (DRDO) Haldwani in India to standardize cultivation of underground/root vegetable crops such as Radish, Beetroot, Turnip and Colocasia under simple hydroponic system- NFT (Nutrient Film Technology). The efforts are ongoing to make the technology cost effective and user's friendly (Agarwal et al., 2021a&b).

Table 1: Performance of underground vegetables under hydroponics system at DIBER, DRDO, Haldwani

Name of crop	Number of leaves/ plant	Number of cormslets per plant	Number of days to harvest	Biological Yield (g/plant)
Radish var. Red globe	11.20	-	48.0	217.33
Radish var. Chinese Pink	10.00	-	52.0	114.67
Turnip var.PTWG	10.67	-	59.0	88.67
Turnip var Cherry Red	14.67		65.0	228.67
Beet root var Agro	15.00	-	70.0	261.0
Colocasia	5.80	12.2	90.0	652.32

DIBER, DRDO has successfully grown and standardized following vegetable crops under hydroponics. The productivity of these crops is as under-

S. No.	Crop	Productivity (kg/sq m)
Leafy vegetables		
1	Spinach (*Spinacea oleracea*)	18-22
2	Lettuce (*Lectuca sativa*)	8-12
3	Lahi (*Brassica oleracea* var *rapa*)	8-12
4	Cabbage (*Brassica oleracea* var *capitata*)	14-18
5	Knol-khol (*Brassica oleracea* var *gongylodes*)	12-16
6	Packchoi (*Brassica rapa* subsp. *chinensis*)	8-10
7	Broccoli (*Brassica oleracea* var *itlaica*)	5-8
8	Cauliflower (*Brassica oleracea* var *botrytis*)	
Fruit vegetables		
1	Tomato (*Solanum lycopersicum*)	10-20
2	Cherry tomato (*Solanum lycopersicum*)	3-5
3	Capsicum (*Solanum annuum*)	2-4
4	Chilli (*Solanum annuum, S. frutescens*)	2-4
5	Cucumber (*Cucumis sativus*)	6-8
6	Bottlegourd (*Lagenaria siceraria*)	4-6
7	Smoothgourd (*Luffa cylendrica*)	3-4
8	Summer squash (*Cucurbita pepo*)	8-10
9	Beans (*Phaseolus vulgaris*)	2-3
Underground/root vegetables		
1	Round radish (*Raphanus sativus*)	2-3.5
2	Turnip (*Brassica rapa* subsp. rapa)	7-10
3	Beet root (*Beta vulgaris*)	3-6
4	Colocasia (*Colocasia esculenta*)	8-10

S. No.	Crop	Productivity (kg/sq m)
Medicinal herbs		
1	Mint (*Mentha* sp.)	5-7
2	Basil (*Ocimum* sp.)	1-2
3	Oregano (*Origanum vulgare*)	2-3
4	Coriander (*Coriandrumsativum*)	3-5
5	Celery (*Apium graveolens*)	6-8

Way Ahead

Although hydroponics is a simple technology but it involves challenges in terms of maintaining nutrient balance of growing media, buffering capacity of the system and substituting nutrient deficiencies in growing media. Therefore, support of computerized programming, automation and artificial intelligent system are required to monitor the system requirement and communicate the requirement in real time for correction for commercial units. The main challenge in taking this technology to masses is requirement of skill and knowledge of system regulation. Although artificial intelligence (AI) has supported the automation of hydroponics unit but the technology is complicated and costly. To make this technology, a technology of every household more customization with user friendly is required with inclusion of hybrid power system like use of solar power or auxiliary power to make it practically acceptable in off-grid areas.

Hydroponics is a versatile technology, appropriate for both village and backyard production systems to high-tech space stations. Hydroponic technology can be an efficient mean for food production from extreme environmental ecosystems such as deserts, mountainous regions, or arctic communities.Hydroponics has shown its promise for the future space program. NASA has put extensive hydroponics research plans in place, which will benefit current space exploration, as well as future, long-term colonization of Mars or the Moon.

References

FAO Report (2009). How to feed the world in 2050. http://www.fao.org/fileadmin/templates/wsfs/docs/expert_paper/How_to_Feed_the_World_in_2050.pdf

Pradhan, B. and Deo. B.(2019), Soil-less farming- the next generation green revolution. *Current Science*, **116** (5): 728-732

NAAS, Vertical farming. Policy paper No. 89, 2019, National Academy of Agricultural Sciences, New Delhi, p20.

Pant, T., Agarwal, A., Bhoj, A. S., Joshi, R. P., Prakash, O. and Dwivedi, S.K. (2018). Vegetable cultivation under hydroponics in Himalayas- challenges and opportunities. *Defence Life Science J.*, **3** (2):111-115.

Agarwal, A., Prakash, O., Sahay, D., Arya, S., Dwivedi, S., K. and Bala, M.(2019a). Performance of lettuce under different soil-less cultures. *Progressive Horticulture*, 2019, **51** (1): 81-84.

Agarwal, A., Prakash, O., Sahay, D. and Bala, M. (2019b). Innovative horticulture: hydroponics (soil-less cultivation) for urban farming. *New Age Protected Cultivation*, **5** (2): 38-40.

Agarwal A., Prakash O., Sahay D. and Bala M. (2021a). Cultivation of *Colocasia* and underground vegetables under Hydroponics- A success report from DIBER. Academia Letters, Article 2811. https://doi.org/10.20935/AL2811

Agarwal, A.; Prakash, O.; Sahay, D. and Bala, M., (2021b). Performance of Spinach under hydroponics and effect on nutritional quality parameters. Research & Reviews-*J. Bot. Sci.*, **10:**107-109.

Chapter 14

Provisions of Plant Variety Protection in India

Uma Kant Dubey
Deputy-Registrar
Protection of Plant Varieties and Farmers' Rights Authority
Government of India, S-2, A-Block, NASC Complex,Opposite Todapur Village,
New Delhi 110012

Plant varieties are vital for the survival of human beings and livestock as they serve as source of food, fiber, fodder, forest and raw materials for domestic households and industries in any society. Even before the science of genetics was discovered in late 19th Century, the scientific foundation of plant breeding science and techniques, since human beings began cultivating plants for a purpose of harvesting its product for survival, the farmers have practiced plant breeding improving the crop generation after generation by evaluating the performance and expression of the plant population they are growing. The farming community and forest dwellers, nurtured existing diversity as is how is to select for better plant type and yield from within the available adapted naturalized plant varieties such as land races to produce plants of use since millennia. As the science of genetics and plant breeding moved forward, formally the trained plant breeders collected genetic diversity through exploration in biodiversity rich areas, used different toolsfor increasing values for cultivation. It is the plant breeders whose ingenious skill to detect variability, work out the genetics behind such variability and accordingly strategies the breeding methods enable development of new plant varieties, different and better than the previously existing ones. Therefore, the Plant Breeders' Right is one of the most crucial intellectual property rights (IPR) in any society as food is basic need for the society. These rights granted to the plant breeders are therefore a means to recognize their contributions that results in the creation of an ecosystem that sustains innovation continuously. Such a system can establish a mechanism to exert responsible stewardship over germplasm, support long term research and development that would be crucial to enhance farm productivity in the form to incentivisation of plant breeding in public and private sectors while promoting the industry to competitively provide high quality seeds as well as planting materials to farmers.

Although ancient Indian history reveals records of naming of a plant variety by either the individual or commune or village native to the variety, in legal terms, the first formal enforcement of Intellectual Property Rights (IPRs) in plants used in agriculture came about in the form of the "Plant Patent Act, 1930" in USA which was followed in Europe by the formation of the Union Internationale pour la Protection des Obtentions Végétales (UPOV) or the International Union for the Protection of New Varieties of Plants in 1961 at Europe which was subsequently revised in 1972, 1978 and 1991. Presently, there are 78 member states as part of the UPOV convention including regional associations like EU, OAPI. India has not acceded to the UPOV Convention but is listed in the category of observer.

Plant variety protection through Plant Breeder's Rights (PBR) was brought into major focus by the General Agreement on Tariffs and Trade (GATT), a multilateral instrument governing international trade. GATT negotiations in Uruguay Round led to the establishment of World Trade Organization (WTO) in 1995. Article 27.3(b) of the Agreement on Trade Related Aspects of Intellectual Property Rights (TRIPs) under WTO, provides that members shall provide protection of plant varieties, either by patents or by an effective sui generis system or by any combination thereof.

The Government of India enacted the Protection of Plant Varieties and Farmers' Rights (PPVFR) Act in 2001 (53 of 2001) to provide for the establishment of an effective sui generis system for protection of plant varieties, the rights of farmers and plant breeders and to encourage the development of the new plant varieties of economic importance.

PPVFR Rules were notified on 12 September, 2003 and amended from time to time, thereafter. Subsequently, for the purposes of the Act, the Government of India having exercised the powers conferred under the section 3 (1) of this Act, established the Protection of Plant Varieties and Farmers' Rights Authority on 11 November, 2005, vide Gazette notification S.O. 1588(E).

Objectives of the PPVFR Act, 2001

The following are the objectives of the Act:

- To establish an effective system for protection of plant varieties, the rights of farmers and plant breeders and to encourage the development of new varieties of plants.
- To recognize and protect the rights of the farmers in respect of their contribution made at any time in conserving, improving and making available plant genetic resources for the development of new plant varieties.
- To protect plant breeders' rights to stimulate investment for research and development both in the public and private sector for development of new plant varieties.
- To facilitate the growth of seed industry in the country that will ensure the availability of high-quality seeds and planting material to the farmers.

Salient Features of the Act

The PPVFR Act, 2001 is based on the compliance under TRIPS Agreement for a member country to protect the rights of plant breeders through sui generis system by enactment by legislation if they would not opt for UPOV (International Union for Protection of New Varieties of Plants, Geneva) Convention, and is also unique in the sense that it concurrently recognizes the rights of plant breeders, farmers (including their right as plant breeders), farming communities and researchers who breed new varieties as well as those already bred and existing prior to protection (extant). It confers exclusive rights upon the breeder or his successor, his agent or licensee, to produce, sell, market, distribute, import or export of the registered variety. As far as farmers' rights are concerned.The Act recognizes a farmer as cultivator, conserver and breeder and provides that the farmers variety can also be registered. Further, the Act provides for compulsory license of a registered variety, if the seeds/propagating material is not available to the public at a reasonable price or quantity. Any person or group of persons or any organization can also claim for benefit sharing, if the plant genetic material belonging to them is used in the development of a registered variety. The researchers are conferred the right to use any registered variety for conducting experiment or research and the use of a variety by any person as an initial source of variety for the purpose of development of other varieties. India is a pioneer country where a national legislation has been enacted to establish and secure Farmers' Rights. The Act also recognizes the past, present and future contributions of the farming communities and provides an opportunity for the award to farming communities/farmers for their contributions in agro-biodiversity conservation.

Plant Varieties Registration

The PPVFR Authority has finalized the Distinctiveness, Uniformity and Stability (DUS) test guidelines for registration of 183 crop species. The Authority has issued 5126 certificates of registration for plant varieties (under new, extant and farmers' variety category) till date on March 23, 2023. To facilitate more applications seeking plant varieties protection from different stakeholders, the Authority regularly organizes/supports awareness and capacity building programmes.

The PPVFR Authority has also established network of DUS test centers across the country under the Central Sector Scheme for the implementation of PPVFR Act, 2001, to verify the claims of candidate varieties by applicants, maintenance breeding, multiplication of reference/example varieties/ the varieties notified under section 5 of the Seeds Act, 1966, and generation of database for varietal characteristics as per crop specific DUS (Distinctiveness, Uniformity and Stability) test guidelines. In addition, DUS tests for the candidate varieties are being conducted at crop specific centers. The data recorded as per the DUS test guidelines is submitted by these centers to Authority for further analysis. The Authority, in consultation with the ICAR institutes and SAUs has identified potential crop species of economic importance and supports projects for the development of the DUS guidelines. The Authority has established its National Gene Bank, field gene banks across the country. It regularly publishes Plant Variety Journal of India and maintains the National Register of Plant Varieties at Headquarters and also its branch offices.

Categories of Protection of Plant Varieties and Registration

The process of granting protection to a plant variety through plant breeders' rights as enshrined in the Act is based on a broad principle of internationally recognized system of DUS and novelty for a new variety. Any person can apply for registration in any of the following:

- New variety of such genera and species as specified under section 29(2) of the Act.
- Extant variety (To a limited period after the species is notified (in the case of new and VCK varieties) as announced time to time by the Authority)
  - Notified under section 5 of Seeds Act, 1966,
  - Variety of common knowledge (VCK),
  - Farmers' variety
  - Traditionally cultivated and evolved by the farmers in their fields
  - Wild relative or landrace of a variety about which the farmers possess common knowledge.
- Essentially derived variety (EDV)

A variety predominantly derived from an initial variety, or from a variety that itself is predominantly derived from such initial variety, while retaining the expression of the essential characteristics that result from the genotype or combination of genotypes of such initial variety.

Farmers' Rights

The Act, provides following rights to the farmers:

- **Right to register their varieties:** A farmer who has bred or developed a new variety is entitled for registration and other protection as a plant breeder under this Act.
- **Right on seed:** A farmer is entitled to save, use, sow, re-sow, exchange, share or sell his farm produce including seed of a variety protected under this Act; provided that the farmer is not entitled to sell branded seed of a variety protected under this Act.
- **Right for reward and recognition:** A farmer or community of farmers engaged in the conservation and improvement of plant genetic resources (landraces and wild relatives of economic plants) are entitled for the Plant Genome Saviour Award, Reward and Recognition, provided their plant material has been used in development of varieties registrable under this Act.
- **Protection of innocent infringement:** If a farmer being, infringed according to section 65 of PPVFR Act, 2001, can prove before court that he or she was not aware of the existence of such rights at the time of infringement; he or she will not be charged.

- **Fee Exemption:** A farmer or group of farmers are not liable to pay any fee payable for inspection of any document or for obtaining a copy of any decision or order or document. Farmers have the privilege of being completely exempted from payment of any kind of fees or other payments that are normally payable for variety registration; testing of varieties and other services rendered by the PPVFRA; as well as for legal proceedings related to infringement or other cases in courts, etc.
- **Reasonable Seed Price:** Farmers have the right to access seed of registered varieties at a reasonable and remunerative price. When this condition is not met, the breeder's exclusive right over the variety can be revoked under the provision concerning compulsory licensing, and the breeder is obligated to license the seed production, distribution and sales of the variety to any competent person. Most of the laws for plant varieties protection have provisions on compulsory licensing of protected varieties to ensure adequate seed supply to farmers at reasonable price.
- **Authorization of farmers' variety:** In the development of an essentially derived variety from a farmer's variety and its commercialization. The authorization should be given with the consent of the farmer or group of farmers who have contributed in the conservation or development of such a variety. Such a process can allow farmers to negotiate the terms of authorization with the breeders, which may include royalties, benefit-sharing etc.
- **Right for compensation:** When any propagating material of a variety registered under this Act has been sold to a farmer, the breeder of such variety should disclose the expected performance under given conditions. If the propagating material fails to provide such performance under such given conditions, the farmer can claim compensation before the Authority. The Authority would then notify the breeder of the variety the issue and after providing him an opportunity to file an opposition, may direct the breeder to pa y compensation to the farmer as it deems fit.

Plant Breeders' Rights

Registration gives exclusive rights to produce, sell, market, export or imports the variety along with its denomination. This right is subject to farmers' rights that farmers can use seeds of registered varieties in an unbranded manner.

Breeders' Right is one of the pivotal provisions of this Act with far reaching implications in the context of Indian agriculture and global scenario. The breeder also enjoys provisional protection of his/her variety against any abusive act committed by any third party during the period between filing of application for registration and the final decision taken by the Authority. Similarly, researcher's rights are also granted. However, for repeated use of a registered variety as an initial source of variety for the purpose of developing a new variety, the authorization of the breeder of the registered variety is necessary.

Extent and Nature of Field-Testing (Dus Testing) of Varieties

The application is processed depending on the category of the variety claimed for registration, accordingly to which the applicant is required to deposit DUS test, registration and any other fees, as may be required. After receipt of necessary fees and seeds and to a satisfactory examination of the application at the Plant Varieties Registry, the Registrar shall send the variety to crop specific centers for conducting DUS test. The period of DUS testing is as follows:

- New Varieties: Two similar crop seasons at two locations.
- Farmers' Variety and VCK: One crop season at two locations.
- Extant variety notified under section 5 of Seeds Act, 1966: No DUS testing is conducted but variety is processed by an EVRC Committee which recommends for registration.
- EDV: DUS testing is not mandatory but field test is conducted to ascertain DUS criteria.

After the receipt of DUS test result, the application is processed and distinctiveness is ascertained through DUS test and comparison across the database. Subsequently, the passport data of the variety is published in the Plant Varieties Journal of India.

The application is advertised in Plant Variety Journal of India inviting pre-grant opposition within a period of three months from the date of publications. If no opposition is filed or if opposition filed is rejected, the variety proceeds for registration. The period of protection is as follows:

Protection Period in Different Types of Crops

A total of 183 crop species are presently eligible for protection. The total period of protection for field crops is of 15 years with 6 years of protection at the time of registration renewable to next 9 years, whereas that of trees and vines is for 18 years with 9 years of protection at the time of registration renewable to next 9 years. The extant varieties notified are given a protection for 15 or 18 years for field crops or trees and vines respectively, from date of notification under Seeds Act, 1966.

Award/Rewards to Farmers'/Farming Communities

Section 45(2) of the Act reads with Rules 70 (2) (a) of PPVFR Rules, 2003 provides for support and reward, to farmers, communities of farmers, particularly the tribal and rural communities engaged in conservation, improvement and preservation of plant genetic resources of economic plants and their wild relatives, particularly in areas identified as agro-biodiversity hotspots from National Gene Fund. To operationalize these provisions, Plant Genome Savior Community Award was instituted in 2009–10. A maximum of five such awards can be conferred annually. Along with this, ten farmers are conferred the Plant Genome Saviour Farmer Reward and twenty farmers are conferred Plant Genome Saviour Farmer Recognition certificates. The details of the awards conferred are mentioned in Table 1. The selection of awardees is made by a committee of experts/ scientists headed by an eminent scientist/ subject matter specialist.

Table 1: Details of the Plant Genome Savior awards, rewards and recognition:

Award	Details	Application
Plant Genome Saviour Community Awards	Five farming communities are awarded each year. Each award includes a citation, a memento and Rs. 10 Lakh	Advertisement for these awards is published in the National dailies and on the Authority website: (http://www.plantauthority.gov.in/forms.htm) The applications should be forwarded by Chairperson/ Secretary of the concerned Panchayat Biodiversity Management Committee or Concerned District Agricultural Officer or Director of Research of Concerned State Agriculture University or Concerned District Tribal Development Officer.
Plant Genome Saviour Farmers' Rewards	Ten farmers are rewarded every year. Each reward includes a citation, a memento and cash of Rs. 1.5 Lakh	
Plant Genome Saviour Farmers' Recognitions	Twenty farmers are rewarded every year. Each reward includes a citation, a memento and cash of Rs. 1 Lakh	

References

Srinivas, C. S. (2003), "Exploring the Feasibility of Farmers' Rights," *Development Policy Review*, vol. 21, no. 4, pp. 419–47.

Srinivasan, C. (2005), "International Trends in Plant Variety Protection," *Journal of Agricultural and Development Economics*, vol. 2, no. 2, pp. 182–220.

Swaminathan, M. S. (1998), "Farmers' Rights and Plant Genetic Resources," *Biotechnology and Development Monitor*, vol. 36, pp. 6–9.

The International Treaty on Plant Genetic Resources for Food and Agriculture, adopted by the 31st Session of Conference of the Food and Agriculture Organization, Rome, 3 November 2001, entry into force on 29 June 2004.

Rangheka. D., 2000. Intellectual property rights Agriculture: An Analysis of the Economics Impact of Plant Breeders Rights.

Nijar, G.S., Sui Generis Law for Plant Varieties: Preserving the Knowledge and Creativity of Traditional Plant Breeders — A Third World View.

Bala Ravi, S. (2004), Manual on Farmers' Rights, M. S. Swaminathan Research Foundation, Chennai.

Website of Protection of Plant Varieties and Farmers' Rights Authority, www.plantauthority.gov.in

Protection of Plant Varieties and Farmers Rights Act, 2001.

Chapter 15

Organic Pest Management Technologies in Vegetable Crops

Prof. (Dr.) N.K. Bajpai
Director Extension
Banda University of Agriculture & Technology, Banda (U.P.)

Organic farming is an old age practice in India and it is one of the several approaches found to meet the objectives of sustainable agriculture. To meet the food, fibre, fuel, fodder and other needs of the burgeoning population in India, the productivity of agricultural land and soil health needs to be improved and it is of utmost importance when the cultivable land resource is shrinking day by day. Though the green revolution in India helped to overcome acute food crisis to a great extent, but the price paid to achieve this satisfaction is beyond the imagination. The adverse effects of modern agriculture resulted in paradigm shift and interest to adopt organic farming globally. Organic farming is a system devoid of the use of any chemical or genetically modified inputs, in which the biological potential of the soil, organic sources and underground water resources are conserved and protected by adopting suitable cropping pattern including agro-forestry and methods of organic replenishment. Organic products are grown under an agricultural practice where chemical fertilizers and pesticides are not used hence it is an eco-friendly, sustainable and sociologically acceptable. This method of farming preserves the regenerative potential of soil, maintains a good plant nutrition and does not harm the environment. The Government of India has initiated the National Programme for Organic Production (NPOP) in 2001. The programme involves the accreditation programme for certification agencies, norms for organic production, promotion of organic farming etc. Under this programme different states e.g., Uttaranchal, Karnataka, Madhya Pradesh, Maharashtra, Gujarat, Rajasthan, Tamil Nadu, Kerala, Nagaland, Mizoram, Sikkim have been promoted for organic farming.

The vegetable crops play a major role in alleviating the malnutrition problem as they are rich source of minerals,vitamins, fibre and contain a fair amount of protein as well as carbohydrates. In addition to local market demand, vegetables have the potential for both domestic and export market. The area, production and productivity under vegetable crops of our country during 1950-51 was merely 2.84 million hectares, 16.5 million tonnes and 5.8 t/ha, respectively and now there has a phenomenal increase in area (3.63 folds), production (10.62 folds) and productivity (2.93 folds) of vegetables during the last seven decades. Presently, vegetables contribute 41.32 % area and 59.32 % production in total horticultural crops in India. As per second advance estimates of National Horticulture Board for the year 2016-17 the total area under vegetables are 10.31 million hectares with a production of 175.19 million tonnes and productivity of 17 t/ha. Although India is the second largest producer of vegetables next only to China in World, the productivity of different vegetables in our country is comparatively lower than the World's average productivity.

Major Benefits of Organic Farming

1. Offers one of the most sustainable farming systems with recurring benefits to long-term soil health.
2. Helps in maintaining environmental health by reducing the level of pollution.
3. Provides a lasting stability in production by importing better resistance against various biotic and abiotic stresses.
4. Reduces human & animal hazards by reducing the level of pesticide residue in the product.
5. Ensures the optimum utilization of natural resources for conserving them for future generation.
6. Healthier and nutritionally superior food for man and animal thus improves the quality of produce.
7. Fetches a premium price of 10%-50% over conventional products.

Why Need of Organic Farming of Vegetable Crops in India

- Most of the vegetable crops are eaten fresh therefore, any contamination may lead to various kinds of health hazards mainly due to pesticides.
- Decrease in land productivity due to increasing use of chemical fertilizers.
- The increased cost of production in conventional farming due to more dependence on chemicals in terms of synthetic fertilizers and pesticides.
- High environmental pollution.
- High demand due to social awareness.

Status of Insect-pests of Vegetable Crops in India

Insect-pests are the key challenges in cultivation of crops since the beginning of cultivation practices. They cause direct damage by feeding root, shoot, leaves, flowers and fruits of the crop thus by reducing the crop yield (Table 1).

Table 1: Yield losses caused by major insect-pests in vegetable crops in India

Crop/Pest	Yield losses (%)
Tomato	
Fruit borer (*Helicoverpa armigera*)	24-65
Brinjal	
Fruit and shoot borer (*Leucinodes orbonalis*)	11-93
Chilli	
Thrips (*Scirothrips dorsalis*)	12-90
Mites (*Polyphagotarsonemus latus*)	34
Okra	
Fruit borer (*Helicoverpa armigera*)	22
Leafhopper (*Amrasca biguttula biguttula*)	54-66
Whitefly (*Bemisia tabaci*)	54
Shoot and fruit borer (*Eariasvittella*)	23-54
Cabbage	
Diamond backed moth (*Plutella xylostella*)	17-99
Cabbage caterpillar (*Peiris brassicae*)	69
Cabbage leaf webber (*Crocidolomia binotalis*)	28-51
Cabbage borer (*Hellula undalis*)	30-58
Cucurbits	
Fruitfly (*Bactrocera cucurbitae*)	
Bitter Gourd	60-80
Cucumber	20-39
Ivy Gourd	63
Muskmelon	76-100
Snake gourd	63
Sponge gourd	50

Source: Dhillon et al., 2005; Ghosalet al., 2012; Raju et al., 2007; Satpathy et al., 2005; Shivlinga Swamy et al., 2002; Singh et al., 2007.

The average crop loss caused by different insect-pets is estimated 30-40 per cent. In addition, many of the insect-pests cause indirect damage and act as vectors of several viral diseases of different crops. Now the management of insect-pests is more difficult because of newer emerging problems e.g., resistance to pesticides, secondary pest outbreak, elimination of natural enemies and new pest threats etc. Insect-pests like serpentine leaf miner in tomato, gall midge in brinjal, stem fly in okra, leaf hoppers in bittergourd and others are posing a threat in cultivation of respective crops in different regions of the country. Andhra Pradesh and Karnataka states are facing the regular problems of gall midge in chilli, capsicum and brinjal. Similar is the case with other insect-pests e.g., head borer in cabbage and red spider mite in okra, brinjal, cowpea and Indian bean, etc. (Table 2).

Table 2: Changing host preference of insect-pests of different vegetable crops in India

Insect-Pest	Old Host crops	New Host crops	References
Serpentine leaf miner (*Liriomyza trifolii*)	Tomato	Brinjal, Cow pea, French bean, Squash, Leafy vegetables, cucurbits	Durairaj*et al.*, 2007
Spiralling whiteflies (*Aleurodicus macfarlanei*)	Guava, Citrus	Okra	Puri*et al.*, 2000
Mealy bug (*Coccidohystrix insolita, Phenacoccussolanpsis*)	Cotton, Jute	Brinjal, Tomato, chilli, okra	Jhala*et al.*, 2008
Hadda beetle (*Henosepilachna vigitioctopuntata*)	Brinjal	Bitter gourd	Rajapaske *et al.*, 2005
Fruit borer (*Helicoverpa armigera*)	Gram, Cotton, Tomato, Cabbage	Peas, Chilli, Brinjal, Okra	Puri*et al.*, 2000
Cabbage butterfly (*Pieris brasicae*)	Cabbage, cauliflower & Mustard	KnolKhol, Radish	Puri*et al.*, 2000
Gall midge (*Asphondylia* sp.)	Brinjal	Chilli and Capsicum	David, 2006
Head borer (*Hellula undalis*)	Mustard, Radish	Cabbage	Sivaprasagam, 1997
Stem fly (*Ophiomyiap haseoli*)	Beans	Okra	Patil&Jamadgni, 2008
Red spider mite (*Tetranychus* sp.)	Greenhouse crops	Okra, brinjal, cowpea, Indian bean	Mahto& Yadav, 2009

Modalities of Pest Management in Organic Farming

Insect-pest management in organic farming is conceptually the same as integrated pest management elsewhere. However, due to the limitations and restrictions in organic farming, the techniques and strategies used tend to focus on augmenting natural processes and hence focus on tactics that produce longer term effects. The management practices include cultural control methods such as use of tolerant/ resistant varieties (Table 3), soil solarization, alteration in date of sowing/planting, intercropping (Table 4) etc. Mechanical control through chromatic traps, mechanical traps, light traps, sticky traps and pheromone traps may give an added advantage. In addition to organic pesticides, semio-chemicals (behavior-modifying compounds) and biopesticides (microbial control agents including entomopathogenicfungi, bacteria and viruses, and botanical biopesticides) are increasing in importance in organic farming. There are a limited number of organic pesticides available and most have short persistence on plants and in the environment. A primary management tactic is the manipulation of natural enemies: releases of commercially reared predators and parasitoids and occasional augmentation by providing alternative host plants and push-pull methods of directing pests away from economic crops and into areas where they are more vulnerable to management.

Table 3: Tolerant varieties of important vegetables in India

Crop	Pest	Varieties
Tomato	Fruit borer (*H. armigera*)	ArkaVikash, Pusa Gaurav, Pusa Early Dwarf, Punjab Keshri, Punjab Chhuhara, Pant Bahar, Azad
Brinjal	Shoot and fruit borer (*L. orbonalis*), Aphid, jassid, thrips, whitefly	SM 17-4, PBR 129-5 Punjab Barsati, ARV 2-C, Pusa Purple Round, Punjab Neelam, Kalyanpur-2, Punjab Chamkila, Gote-2, PBR-91, GB-1, GB-6
Cabbage	Aphid (*Brevicoryne brassicae*)	All season, Red Drum Head, Sure Head, Express Mail
Cauliflower	Stem borer (*Hellula undalis*)	Early Patna, EMS-3, KW-5, KW-8, Kathmandu Local
Okra	Jassid (*Amrasca biguttula biguttula*)	IC-7194, IC-13999 New Selection, Punjab Padmini
Onion	Thrips (*Thrips tabaci*)	PBR-2, PBR-6, ArkaNiketan, PusaRatnar, PBR-4, PBR-5, PBR-6
Round gourd, Pumpkin, Bitter gourd	Fruit fly (*B. cucurbitae*)	ArkaTinda
		Arka Suryamukhi
		Hisar-II

Source: Bhardwaj, 2010; Dhandapaniet al., 2003

Table 4: Pest management in vegetable crops through intercropping

Insect-pests	Crops/Intercrops	Source
Diamond Back Moth	Cabbage + Carrot	Buranday and Raros, 1973; Bach and Tabashnik, 1990
	Cabbage + Tomato	Srinivasan and Veeresh, 1991
	Cabbage + Indian Mustard	Srinivasan and Moorthy, 1992
	Cabbage + Chinese Cabbage	Satpathy *et al.*, 2010
Root fly	Cabbage + French bean	Hofsvang, 1991
Shoot and fruit borer	Brinjal + Coriander/Fennel	Khorsheduzzaman *et al.*, 1997; Satpathy and Mishra, 2011
Fruit fly	Bitter gourd + Maize	Shooker *et al.*, 2006
Yellow vein mosaic	Okra + Corn	Adhikary *et al.*, 2015
Flea beetle	Broccoli + Faba bean	Garcia and Altieri, 1992

Trap cropping is a useful strategy in the management of several pests in various cropping systems. Trap cropping is not a clear-cut solution to all insect-pest problems, but it can be an effective tool against insect-pests. However, it is not essential that all insects can be controlled with trap crops, overall trap crops appear to be a good alternative pest management strategy. The important trap crops commonly used in pest management include, Indian mustard used as trap crop in cabbage to attract diamondback moth, aphid and leaf Webber. Two rows of Indian mustard after 25 rows of cabbage (1st row is sown 15 days prior to cabbage planting and 2nd row 25 days after planting). Sesamum is commonly being used as trap crop to attract Diamondback moth in both cabbage and cauliflower. Two rows of sesamum for every 25 rows of cabbage or cauliflower can be planted to trap the pest. In tomato marigold or cucumber is commonly used as trap crop for every 15 rows of the main crop to attract tomato fruit borer. In case of field beans, chrysanthemum acts as a trap crop against leaf miner. Marigold is a potential trap crop in potato against nematodes. Bihar hairy caterpillar in cowpea can be trapped by planting Gingelly. Maize plants with bait spray attract and kill the adults of fruit fly when bitter gourd is sown as border crop. Planting of castor as trap crop deviates the population of *Spodoptera litura* from cowpea. Trap cropping has indicated a great benefit interms of economic returns on an average of 10-30 per cent increase in net profits mainly resulting from reduced insecticide use and pest attack.

An organic farming system offers many opportunities for enhancing biological control factors. Since most of the organic farms apply little or no harsh pesticides, it is possible to build up large numbers of beneficial parasites and predators that help in managing insect-pests of different crops. Natural enemies, either generalists or specialists, can be utilized to manage the insect-pests effectively. Lady bird beetles are considered generalists because they feed on different kinds of prey; whereas, egg parasitoid *Trichogramma* are microscopic wasps specialized to parasitize eggs of moths. Lacewings, lady bird beetles, syrphid fly larvae, and parasitized aphid mummies are almost always present in aphid colonies. Fungus infected adult flies are often common following periods of high humidity. These natural controls are important and need to be conserved when making pest management decisions.

Inundative release of *T. brasiliensis* @ 240000/ha is also recommended for control of fruit borers on okra and tomato (Jalali *et al.*, 2005). Six releases of *T. brasiliensis* at weekly interval @ 40000/ha the first release coinciding with 50% flowering and thereafter at weekly interval in tomato is recommended. *Chrysoperla zastrowiarabica* is an effective predator for control of white fly, aphid, jassid and eggs of some lepidopterous borers when the first instar larvae are released @ 50000/ha (Khulbe *et al.*, 2005). *Cotesia plutellae* and *Diadegma semiclausum* can be incorporated in to biological pest management of diamondback moth as larval parasitoids (Saucke *et al.*, 2000). Microbial biopesticides are eco-friendly and compatible with natural enemies existing in the ecosystem. It has been widely applied against *Plutella xylostella, Earias vitella* and *H. armigera.* Because of host specificity, use of Nuclear Polyhedrosis Virus (NPV) is limited to *H. armigera* and *Spodoptera litura.* Lack of commercial production has suppressed its popularity and is used in limited areas. Some botanical based biopesticides are also recommended against insect-pests of vegetable crops (Table 5).

Table 5: Botanicals used against different insect-pests of vegetable crops

Insect-pests	Plant products	Mode of action	References
Leucinodes orbonalis	Oil (2%) of *Pongamia pinnata, Madhuca indica* and *A. indica*	Contact	Srinivasan *et al.* (1998)
	Neem Azal (0.03%), Neem oil (0.3%)	Antifeedant & contact	Peter & Govindarajulu (1994)
Epilachna vigintiocto-punctata	Leaf extract (1%) of *Lantana camara*	Antifeedant	Mehta *et al.* (1995)
	Leaf extract (6%) of *Ageratum haustonianum, Melia azedarach*	Contact	Thakur & Mehta (2004)
Whitefly	Neemal® **(%0.5),** Repelin® (Neem based formulation)	Mortality	Somasekharaet *al.* (1997)
Pieris brassicae	Achook, Neem oil and Nimbicidene	IGR, Antifeedant	Dhingraet *al.* (2005)
Helicoverpa armigera	Methanolic of *Vincarosea* and *Callistemon lanceolatus*	IGR, Antifeedant	Halderet *al.* (2009)
Spodoptera litura	Leaf extract of *Persea americara*	Antifeedant& Contact	Wheeler & Islam (2001)
Earias vittella	NSKE (5%)	Contact and ovicidal	Sarode & Gabhane (1998); Sojitra & Patel (1992)
Pest complex of okra	NSKE (5%), Multineem (2.5 lit/ha)	Contact	Misra (2002); Satpathyet *al* (2004)

Insect-pests	Plant products	Mode of action	References
Pest complex of crucifers	Nimbecidine (7.5 ml/lit)	Contact	Ojha& Singh (2003)
Pest complex of tomato	NSKE/Melia seed extract (5%)	Contact	Senguttuvan*et al.* (2005 a&b)
Pest complex of eggplant	Nimbecidine/ Neemgold/Neem Azal (0.1%)	Contact	Srinivasan & Sundarababu (2000)

Products collected from the local farm, animals, plants and micro-organisms and prepared at the farm are allowed for control of pests and diseases. (e.g., Neem Seed Kernel Extract, cow urine spray). Use of genetically engineered organisms and products are prohibited for controlling pests and diseases. Similarly, use of synthetic growth regulators is not permitted. Preparations of few recommended biopesticides for organic farming are given hereunder (Table 6).

Table 6: Recommended biopesticides and their application against insect-pests

Inputs	Methods of Preparation	Application
Panchagavya	It is cow excreta based indigenous nutrient solution. Panchagavya consists of products viz. cow dung, cow urine, milk, curd, jaggery, ghee, banana, Tender coconut and water. When suitably mixed and used, these have miraculous effects. The preparation steps of panchagavya is as follows: 7 kg. cow dung and 1 kg. cow ghee is mixed thoroughly and kept for 3 days. 2. After 3 days, 10 lt. cow urine and 10 lt. water is added, mixed and kept for 15 days with regular mixing both in morning and evening hours. After 15 days the following ingredients are added and mixed Cow milk - 3 liters Cow curd - 2 liters Tender coconut water - 3 liters Jaggery - 3 kg Well ripened banana – 12 nos. Panchagavya is ready after 30 days	3% solution was found to be most effective compared to the higher and lower concentrations investigated. 3 litres of Panchagavya to every 100 litres of water is ideal for all crops.
Lantana leaf extract 10%	Leaves of *Lantana camara* were collected from the nearby area of the farm and 10% aqueous leaf extract is prepared firstly by grinding the leaves and then soaking 100g of grinded leaves in 200 ml. distilled water for 24 hours at a room temperature of 30°C. The aqueous extract was obtained by filtering the mixture (leaf and water) through a Whatman No .42 filter paper and diluted with distilled water to prepare 10% concentration.	The extract is diluted with water @ 10% before spraying. This foliar spray act as insect-pest repellent. It can be sprayed 3-4 times during the crop duration according to pest infestation.

Vermiwash	It is a liquid that is collected after the passage of water through a column of worm action in vermicomposting.	This liquid manure is applied as foliar spray (10% solution) to the plants for better growth and insect-pest and disease management.
Botanicals	These are aqueous or alcohol extracts of different herbs, shrubs or tree.	These extracts are applied as seed treatment before sowing which stimulates plant growth and productivity and act as biopesticide too. Foliar spray and soil drenching were also done in some cases.
Pesto neem	Neem biopesticide is made from cold pressed neem kernels and its active azadirachtin 1500ppm is used as a general insecticide, fungicide and for coating urea for slow release	It is a bio-based pest controller containing 0.5% Azadirachtin and other vital bio-energizers. Application of pestoneem increase resistance to infestation of pest and disease.

Permitted or not Permitted Products for Insect-pest Management in Organic Farming

There are certain products which are environmentally safe and permitted to use in organic farming. However, there are other products which can be used in restricted manner and its use will be regulated by the certification agency (Table 7).

Table 7: List of approved/not approved/restricted products used in organic farming

Plant based extracts	Permitted
Tobacco tea	Not allowed
Sea weeds, sea weed meal, sea weed extracts, sea salt and salty water	Restricted
Chlorides of lime/soda	Restricted
Clay (Bentonite, perlite, vermiculite, zeolite)	Permitted
Light mineral oils	Restricted
Lime Sulphur (calcium polysulphide)	Restricted
Sulphur (as acaricide, repellent)	Restricted
Viral preparations	Permitted

Fungal preparations	Permitted
Bacterial preparations	Permitted
Parasites, predators and sterilized insects	Permitted
Soft soap (potassium soap)	Permitted
Homeopathic and Ayurvedic preparations	Permitted
Herbal and biodynamic preparations	Permitted
Light trap, sticky trap, pheromone trap, mechanical and chromatic trap	Permitted

(Source: APEDA, 2005)

Pest Management Practices in Important Vegetable Crops through Organic Approach

Tomato

For fruit borer management

- Two rows of marigold as trap crop
- Fourteen rows of tomato
- Two releases of *Trichogramma brassiliensis* @ 50,000 parasitized eggs/ha
- first coinciding with flowering initiation and second 10 days later
- spray of *Ha*NPV @ 350 LE /ha twice
- Inundative release of *T. brassiliensis* @ 2,50,000 eggs/ha alone beginning with fruiting could reduce 16% fruit infestation

For leaf miners and whiteflies

- Cultural and biological control as well as sprays of soft insecticidal soaps (one table spoon of soap powder in 4 L of water) and one or two tea spoons of oil plus Azadirachtin (Nimbecidine EC available formulation 0.03% = 300ppm: Dosage @ 5 ml/l of water).
- Monitor the activities of the adult whiteflies by setting up yellow pan traps and sticky traps at 30 cm from the ground. Yellow empty tins or plates smeared with Castor oil can also be kept in the field. Whiteflies get trapped on them, and they can be wiped off every day and the castor oil re-applied. The number of traps needed depends on the intensity of the pest. 50 traps/ha will be optimum.
- Collect and remove whitefly infested leaves from the plants and those which were shed due to the attack of the pest and destroy them.
- Spray a 5% solution of Neem Seed Kernel extracts mixed with neem oil. To prepare enough mixture of the extract to spray an area of 4000 m^2, 3-5 kg of neem kernel is required. Remove the outer seed coat and use only the kernel. If the seeds are fresh, 3 kg of kernel is sufficient. If the seeds are old 5 kg is required.

Brinjal

For shoot and fruit borer

- Installation of plastic funnel traps 100/ha baited with sex pheromone lures from 15-20 days of planting traps at 10 m distance
- Clipping of damaged shoots and early infested fruits at weekly interval
- Need based foliar spray of NSKE (4%) Reduced infestation up to 60- 65% additionally with safer molecules.
- Inundative release of *T. chilonis*
- Development of insecticide compatible egg-parasitoids
- Conservation of *Trathala flavo-orbitalis*

Okra

For shoot and fruit borer, whiteflies and mites

- Many weed plants harbour whiteflies, removal of weed hosts found to reduce both the incidence of whiteflies and associated viral diseases.
- Destruction of infested plant parts during initial stage of attack.
- Conserve natural enemies, parasitic wasps and predatory insects, including lady bird beetles, damsel flies, lacewings and hover fly larvae are important in natural control of pests.
- Neem-based pesticides, namely Achook © (Azadirachtin 15% w/w) and a neem extract (Azadirachtin 0.6 w/w) in the post flowering period at 49, 61, 73 days after sowing can effectively control the pests.
- Regular monitoring by light/pheromone traps (10 traps/acre) for early pest detection.
- Field release of biological control agents (*Trichogramma brasiliensis/T. chilonis*@ 50,000/ha against *Helicoverpa armigera* & *Earias* spp.)

Cole Crops

For DBM management

- Growing paired mustard rows as trap crop with 25 cabbage rows (first row 15 days before, second 25 days later.
- Need based NSKE (5%) spray
- Such practices conserve *Cotesia plutellae*, a dominant natural enemy
- **Fine tuning with using** Chinese cabbage (*Brassica rapa* subsp. *pekinensis*) as trap crop improved the attraction/trapping of DBM larvae 9 times more than mustard

For Spodoptera *management*

- In early cauliflower and cabbage, hand removal and destruction of egg masses
- Three foliar spray of SlNPV @ 300 LE/ha at 10 days' interval.

For Aphid management

- Coriander as border crop
- Spray of entomopathogenic fungal (*Verticillium lecani*) formulation 500g/ha at 10 days' interval

Chilli

For whiteflies, thrips, mites and fruit borer

- Nursery in pots under screen cover over 1.5 months.
- Use of white and silvery plastic mulch for thrips and delayed virus epidemic.
- Application of neem cake @ 100 kg/acre is advisable.
- Use of yellow sticky traps.
- Use of border crops like maize, jowar, etc. as a barrier for sucking insect-pests.
- Change in the agronomic practices to disturb the life cycle of the grub is also found useful.
- Grass can be heaped at different places in the field and the grubs which accumulate in these heaps may be collected in the early morning and destroyed.
- 400 g/acre of *Beauvaria bassiana* may be broadcasted in the field. Application of neem seed kernel extract (NSKE) can be done for control of thrips, aphids and mites.
- Release of larvae of *Chrysoperla carnea*, a bio control agent, once in 15 days is also helpful in controlling thrips and mites.

Cucurbits

For Fruit fly management

- Use of plastic 25 bottle traps/ha baited with cue lure saturated wood blocks (ethanol: cuelure: carbaryl - 8:1:2) prior to flower initiation reduced infestation up to 65-70% along with use of botanicals
- **Fine tuning of this module** with addition of NSKE spray and increased diameter of bottle trap (9cm)

Increasing consciousness about conservation of environment as well as health hazards related to synthetic chemicals and increasing pattern of consumer preferences towards safe and hazard free vegetables resulted in developing interest of growers to look upon organic farming. Lack of knowledge about biocontrol agents and lack of availability of IPM devices and other organic inputs is the major constraints for farming communities to adopt the organic farming successfully. By enhancing the knowledge base of the farmers with different tools and farmer friendly policies, improving the skills of farmers, availability of inputs at various levels can bring a significant change in the status of organic farming in India. As for as insect-pest management is concerned Integrated Pest Management (IPM) is an important component of organic farming. IPM incorporates a number of strategies, including pre-planting practices, to reduce pest populations and ensure a successful harvest. Role of public-private partnership in production, distribution and quality control of different components of IPM such as resistant varieties, natural plant products, biopesticides and natural enemies is imperative, otherwise we will continue to talk of alternative methods of control for another many years. The necessity to obtain an organic certificate arose because most farms converted their chemical farming to organic farming therefore, an agency must ensure that synthetic chemicals are not included in the cultivation of crops. Furthermore, there is also a need of research backup to support the adaptation of organic farming as a means of mitigating climate change.

References

Adhikary, S.;Pandit, M.K;Koundinya, A.V.V.;Bairagi, S. and Das, A. (2015). Examination of system productivity and profitability of baby corn based vegetable intercropping systems. Journal Crop and Weed 11(1): 220-224.

Bach, C.E. and Tabashnik, B.E. (1990). Effects of non-host plant neighbours on population densities and parasitization rates of the diamondback moth (Lepidoptera: Plutellidae). Environ. Entomol. 19: 987-994.

Buranday, R.P. and Raros, R.S. (1973). Effects of cabbage-tomato intercropping on the incidence and oviposition of the Diamondback moth, *Plutellaxylostella*(L.), Philipp. Entomol. 2: 369-374.

Dhandapani, N.U.;Shekhar, R. and Murugan, M., (2003), Bio-intensive pest management (BIPM) in major vegetable crops: An Indian perspective. Food, Agric. Environ., 1:333-339.

Dhillon, M.K.; Singh, R.;Naresh, J.S. and Sharma, H.C. (2015). The melon fruit fly, *Bactroceracucurbitae*: A review of its biology and management. Journal of Insect Science, 5: 40.

Durairaj, C.;Shobanadevi, R.; Suresh, S. and Natrajan S., (2007). A non-chemical method for the management of leafminer*Liriomyzatrifolii* and whitefly *Bemisiatabaci* in brinjal. Acta Horticulture, 752: 527-529.

Garcia, M.A. and Altieri, M.A., (1992). Explaining differences in flea beetle *Phyllotretacruciferae*Goeze densities in simple and mixed broccoli cropping systems as a function of individual behavior. Entomol. Exp. Appl. 62, 201–209.

Ghosal, A.; Chatterjee, M. L. and Manna, D. (2012). Studies on some insecticides with novel mode of action for the management of tomato fruit borer (*Helicoverpaarmigera* Hub.). Journal of Crop and Weed, 8(2): 126-129.

Halder, J.; Srivastava, C.;Dhingra, S.;Dureja,P. and Tanwar, R.S. (2009). Antifeedant activity of methanolic extracts of *Vincaroseaa*nd *Callistemonlanceolatus*against *Helicoverpaarmigera*. Ann. Pl. Protec. Sci. 17 (2): 496-497.

Hofsvang, T.(1991).The influence of intercropping and weeds on the oviposition of the brassica root flies (*Delia radicum*and *D. floralis*). Norwegian Journal of Agricultural Sciences 5(4): 349-356.

Jalali, S.K.; Murthy, K.S.;Venkatesan, T.; Lalitha, Y. and Devi, P.S. (2005). Preliminary testing of some new release methods for egg parasitoid *Trichogramma* spp. Journal of Biological Control 19(2): 99-103.

Jhala, R.C.;Bharpoda, T.M. and Patel, M.G. (2008). *Phenacoccussolenopsis* Tinsley (Hemiptera: Pseudococcidae), the mealybug species recorded first time on cotton and its alternate host plants in Gujarat, India. Uttar Pradesh J. Zoology, 28, 403-406.

Khorsheduzzaman, A.K.M.; Ali, M.I.;Mannan, M.A. and Ahmed, A. (1997). Brinjal-coriander intercropping: An effective IPM component against brinjal shoot and fruit borer, *Leucinodesorbonalis*Guen. (Pyralidae: Lepidoptera). Bangladesh Journal of Entomology 7(1/2): 85-91.

Khulbe, P.;Maurya, R.P. and Khan, M.A. (2005). Biology of *Chrysoperlacarnea* (Stephens) on different host insects. Annals of Plant Protection Sciences 13(2): 351–354.

Mahto, T.P. and Yadav, R.P. (2009). Effect of manuring with oilcake based vermicompost on incidence of red spidermite of Okra crop. Annals of Plant Protection Sciences, 17:40-45.

Mehta, P.K.; Vaidya, D.N. and Kashyap, N.P. (1995). Antifeedant properties of some plant extracts against brinjalhadda beetle, *Henosepilachnavigintioctopunctata* Fab. J. Entomol. Res. 19(2): 147-150.

Misra, H. P. (2002). Field evaluation of some newer insecticides against aphids (*Aphis gossypii*) and jassids(*Amrascabiguttalabiguttala*)on okra. Indian J. Entomol, 64: 80–84.

Ojha, P.C. and Singh, I.P. (2003). Bio-efficacy of some biorational insecticides against insect pest complex of cauliflower. Pestology, 27(4): 23–25.

Peter, C. and Govindarajulu, P. (1994). Efficacy and persistent toxicity of certain new insecticides against brinjal pest complex. Pestology, 18(1): 27–30.

Puri, S.N.; Murthy, K.S. and Sharma, O.P. (2000). Integrated Pest management in vegetables: Issues and strategies. In. Kalloo, G. and Singh, K. (eds.) Emerging scenario in vegetable research and development. Research Periodicals and book publishing house, Texas, USA. pp- 293-303.

Raju, S.V.S.; Bar, U.K.; Uma Shankar and Kumar, S. (2007). Scenario of Infestation and Management of Eggplant Shoot and Fruit Borer, *Leucinodesorbonalis*Guen. in India. Resistant Pest Management Newsletter, 16(2): 14-16.

Sarode, S.V. and Gabhane, A. T. (1998). Effect of neem and insecticidal combinations for okra fruit borer management. *Pest Manage. Hortic. Ecosys*, 4: 54–56.

Satpathy, S.; Kumar, A.; Singh, A.K. and Pandey, P.K. (2005). Chlorfenapyr: A new molecule for diamondback moth (*Plutella xylostella* L.) management in cabbage. Annals of Plant Protection Sciences, 13: 88-90.

Satpathy, S. and Mishra, D.S. (2011). Use of intercrops and antifeedants for management of eggplant shoot and fruit borer *Leucinodes orbonalis* (Lepidoptera: Pyralidae). International Journal of Tropical Insect Science 31 (1-2): 52-58.

Satpathy, S.; Shivalinga Swamy, T.M.; Kumar, A.; Rai, A.B. and Rai, M. (2010). Potentiality of Chinese cabbage (*Brassica rapa* subsp. *pekinensis*) as a trap crop for diamondback moth (*Plutella xylostella*) in cabbage. Indian Journal of Agricultural Sciences 80(3): 238-241.

Satpathy, S.; Rai, S.; De, N. and Singh, A.P. (2004). Effect of insecticides on leaf wet carbon assimilation rate and pest incidence in okra. Indian J. Plant Prot., 32: 22–225.

Senguttuvan, K.; Kuttulam, S.; Manoharan, T. and Gunasekaran, K. (2005a). Investigation of azadirachtin content in*Azadirachta indica*A. Juss ecotypes of Tamil Nadu. Pestology, 29(1): 34–36.

Senguttuvan, K.; Kuttulam, S.; Manoharan, T. and Srinivasan, T. (2005b). Bioefficacy of *Meliadubia* Cav. and neem products against major insect pests of tomato. Pestology, 29(1): 47–50.

Shivalingaswamy, T.M.S.; Satpathy, S. and Banerjee, M.K. (2002). Estimation of crop losses due to insect pests in vegetables, In: Resource Management in plant protection, Vol-I (SaratBabu, B.; Vara Prasad, K.S.; Anita, K.; Rao R.D.; Chakraborty, S.K. and Chandukar, P.S. eds.) Plant Projection association of India, Hyderabad, pp-24-31.

Shooker, P.; Khayrattee, F. and Permalloo, S. (2006). Use of maize as a trap crops for the control of melon fly, *B. cucurbitae* (Diptera:Tephritidae) with GF-120. Bio-control and other control methods (online). Available on:http\\www. fcla. edu/flaEnt/fe87 p354. Pdf.

Singh, S K.; Singh, A.K. and Singh, H.M. (2007). Relative resistance of okra germplasm to shoot and fruit borer, *Earias vittella* Fabr. under field conditions. Journal of Applied Zoological Researches 18(2): 121-123.

Sojitra, I.R. and Patel, J.R. (1992). Effect of plant extract on ovipositional behaviour of spotted bollworm, *Earias vitella* infesting okra (*Hibiscus esculentus*). Indian J. Agric. Sci., 62: 848–849.

Somasekhara, Y. M.; Nateshan, H. M. and Muniyappa, V. (1997). Evaluation of neem products and insecticides against whitefly (*Bemisia tabaci* a vector of tomato leaf curl geminivirus disease. Indian J. Plant Prot., 25: 56–59.

Srinivasan, G.; Babu, P.C.S.; Reddy, P.P.; Kumar, N.K.K. and Verghese, A. (1998) Advances in IPM for horticultural crops. In: Proceedings of the first national symposium on pest management in horticultural crops: environmental implications and thrusts, Bangalore, India, pp 87–93

Srinivasan, K. and Moorthy, P.N.K. (1992). Development and adoption of IPM for major pests of cabbage using Indian mustard as trap crop. In: Diamondback moth and other crucifer pests: Proceedings of 2nd International Workshop, Taiwan, 10-14 December, 1990.

Srinivasan, K. and Veeresh, G.K. (1986). The development and comparison of visual damage threshold for the chemical control of *Plutellaxylostella* and *Crocidolomiabinotalis* on cabbage in India. Insect Sci. Applic. 7 (4): 547-557.

Srinivasan, G. and Sundarababu, P. C. (2000). Comparative efficacy of neem products against brinjal leafhopper, *Amrasca biguttala biguttala. Indian J. Entomol*, 62: 18–23.

Thakur, M. and Mehta, P. K. (2004). Assessment of insecticidal properties of some plants against Hadda beetle,*Henosepilachnavigintioctopunctata*(Fabr.). *Indian J. Entomol*, 66: 20–23.

Wheeler, D.A. and Isman, M.B. (2001). Antifeedant and toxic activity of *Trichilia americana* extract against the larvae of *Spodpteralitura*. Entomol. Exper. & Appl., 98: 9–16.

Chapter 16

Futuristic Technologies in Integrated Pest Management of Apple for Profitability and Sustainability

S.S. Singh[1], S.K.Dwivedi[*], Renu[3], Mukta Nainwal[4] and Rakesh Mer[5]

G.B. Pant University of Agriculture & Technology, Krishi Vigyan Kendra, Matela-263643, Almora, Uttarakhand, (Email: sssindia02@gmail.com)

[*]Scientist "G" (Horticulture), DRDO, New Delhi

Apple (*Malus* x *domestica Borkh.*) holds an important position in the fruit industry of Indian Himalayas as it accounts for about 2.05 % in total world apple production and 2.36 % of the total fruit production of the country. India is the fifth largest producer of Apple in the world with area of 0.308 million hectares and production quantity of about 2.316 million tonnes of which about 70.54 % comes from Jammu and Kashmir, 26.42 % from Himachal Pradesh, 2.66 % from Uttarakhand and 0.30% from Arunachal Pradesh, which however varies from year to year as the crop is highly sensitive to weather conditions (APEDA Agri Exchange, 2021, Bhalla, and Gupta, 1993 and Singh and Lal, 2004). Apple is one of the oldest and commercially important temperate fruit that widely produced an important pome fruit in the world. It is a temperate fruit crop of the North-Western Himalayan region and predominantly grown in Jammu and Kashmir, Himachal Pradesh and Uttarakhand in the NW region of India (Sharma 2018). In India higher intake and careless use of pesticide leads to residual effect on food and water that is indirectly consumed by human and cause health hazards. thus, the pesticides are hazardous in nature with very high cost and also create the resistance with time So, there is need to adopt new strategy for the conservative and utilization of resources available to alter the behavior of insect pest adversely affect the crops for sustainability in agriculture production by reducing the cost of pest management (Gaikwad *et al.*, 2019).

According to the Food and Agriculture Organization of the United Nations (FAO), Integrated Pest Management (IPM) is an attentive approach using new technologies for pest control and ensures implementation of all appropriate measures that reduces the infestation of pest and uses those measures which are economically justified to minimize risks of human health and the environment. IPM promotes the growth of a healthy crop with the minimum possible disruption to agro ecosystems and encourages natural pest control mechanisms" (FAO 2020). Coll and Wajnberg (2017) reported that integration counts all the pests of a particular crop but Integration is about controlling a single or a small number of pest species. IPM also promote sustainable agriculture, minimal use of synthetic pesticide, resolves socio-economic, environmental, and human health challenges. Global pesticide use shows negative inference for farmer livelihoods, biodiversity conservation, and the human right to food (Deguine *et al.*, 2021).

The apple grown in Kashmir famous for its unique taste and delicacy, but fruit growth, yield and quality are directly influenced by several biotic and abiotic factors. Among these factors, insect pests and diseases cause huge economic loss thus the growers invest so much pesticide to cure the damage (Altaf *et al.* 2019) The study shows that more than 90% area of apple in these states is covered under exotic varieties which are no doubt high yielding but most of them are susceptible to various pests and diseases. Besides, Apple is grown by small and marginal farmers and for most of the inputs including pesticides, these farmers depend upon the traders who guide them keeping in view their profits that hinder the transfer of proper technology. Moreover, the Apple growers rarely follow the recommended eco-friendly pest and disease management practices. All these factors result in higher incidence of pests and diseases. Due to the lack of resistant varieties and rootstocks, poor adoption of improved cultural practices, inefficient biocontrol agents and poor availability of biopesticides and neem-based formulations, the integrated Pest Management strategies are dominated by excessive use of pesticides. With the result, the Apple crops are grown under pesticide umbrella. However, excessive and indiscriminate use of pesticides especially in Himachal Pradesh and Jammu & Kashmir, since the last one and half decades have adversely affected the Apple ecosystem, with the result even a slight disturbance leads to outbreak of one or more pests in one zone or the other.

According to an estimate, pests and diseases causing over Rs. 90,000 crore crop losses including Apple per annum and for combating their menace pesticides is the most powerful weapon used in the country. Due to the lack knowledge of insect pest management in India the exotic insect pests are of major concern. Such insects like thrips, leaf rollers and tussock moth cause severe infestation of apple in Jammu and Kashmir and cause huge loss. It depends on the nature of infestation, identification and detection of different stages of apple pests and their distribution in India. (Hussain *et al.*, 2018).In view of the rising export potential of residue free fruits and the crisis situation created by indiscriminate use of pesticides, apple crop required much more attention in terms of IPM input than what it is getting now. With increasing global trade system in agriculture has increase the introduction of severe exotic pests. Due to change in the cropping pattern, indiscriminate use of pesticides, climate change and introduction of input intensive high yielding varieties, hybrids, and cultivars there is an approached shift in pest status. Pest has expanded their host horizon, create resistance against pesticides and become a serious pest. (Rathee, 2018). It is therefore, highly imperative to develop and implement a rational approach with greater reliance on IPM to promote sustainability and to reduce the number of applications of hazardous pesticides in temperate fruits. Apple production requires those strategic adaptation options under changed climatic scenarios to enhance the production .The study stated that apple crop needs adequate chilling for breaking its winter dormancy for a usual spring growth. The traditional apple crop areas are shifting upward in hills as a result less suitable areas transforming in unsuitable and severely cold areas which were earlier not suitable are becoming suitable for apple cultivation. (Singh and Patel, 2017)

Present Status

Apple can be grown at altitudes 1,500-2,700 m. above m.s.l. in the Himalayan range which experience 1,000-1,500 hours of chilling (the no. of hours during which temperature remains at or below 7^0 C during the winter season). The temperature during the growing season is around 21-24^0 C. For optimum growth and fruiting, apple trees need 100-125 cm. of annual rainfall, evenly distributed during the growing season. Excessive rains and fog near the fruit maturity period result in poor fruit quality with improper colour development and fungal spots on its surface. Areas exposed to high velocity of winds are not desirable for apple cultivation. The study stated that Small-scale farms are an important component of agricultural production even in developed economies, social development with conservation of agricultural biodiversity and fulfill the demand of food security. This review details some of the characteristics of the small farm sectors in developed economies and identifies some of the characteristics of small farms and their operators that may favor the implementation of IPM. Knowledge gaps and opportunities are responsible for IPM outreach to small-scale producers. (Grasswitz, 2019)

The apple crop acreage in Indian Himalayas is rapidly increasing and is developing into perennial stretches of monocroppinig systems. Parally this has led to increase pests and diseases incidence and use of chemical pesticides, especially synthetic pyrethroids. All these have reduced the quality of fruits affecting domestic and export market. Introduction of modern fruit crop production practices and increased use of pesticides have no doubt increased the fruit production many fold but has also resulted in changes in pest and diseases scenario over space and time and many new pest and disease problems have emerged. (Gupta, 1999 Singh, and Lal, 2002).The premature leaf fall in apple was first detected in 1995 in some orchards, has now spread to all apple orchards in Himachal Pradesh and can cause up to ninety per cent of defoliation in unprotected orchards. All commercial delicious cultivars are susceptible to this disease. Premature leaf fall in apple has also been reported recently from Kashmir, the neighbouring Uttaranchal and Bhutan. (Sharma, 2003).Recently, two species of the Red Spider mites, European Red mite (*Panonychus ulmi*) and Two –Spotted mite (*Tetranychus urticae*) have emerged as new serious pest problem in apple growing areas of India. The pest has now appeared in an epidemic form in all apple-growing states of India. At present, the European Red mite and Two–Spotted mite occur in 70:30 ratio in the orchard ecosystem (Kakar, 2000). Due to improper orchard management huge number of insect pests attacks the apple crops, some of them are very serious so there is a need to take attentiveness for their control. Due to unawareness of the people the sale of some banned pesticides as Endosulfan, and unlabeled and factitious use of pesticides has incensed these problems to great extent. Injudicious and uncontrolled use of chemical pesticides in agriculture responsible for ecological imbalances and environmental pollution (Georgiadis *et al.*, 2018; Kumari and John, 2019). The excessive and indiscriminate use of chemical pesticides coupled with favourable microclimate due to ignorance of improved cultural practices is mainly responsible for widespread incidence of European red mite in Himachal Pradesh and Jammu & Kashmir. Moreover, Apple dapple (a viroid pathogen) is becoming a serious problem in many orchards situated in Kotkhai and Jubbal areas of Himachal Pradesh.

Intensity of white root rot, collar rot, canker diseases, powdery mildew and wooly aphid has also increased. Similarly codling moth in apple known to be minor pest has become a regular and severe pest of Leh and Kargil districts of Ladakh region. Similarly, Apple mosaic and Apple chlorotic leaf spot which were sporadic disease of apple earlier in Himachal Pradesh have become serious problem. Apple mosaic is caused by a virus and produces mosaic symptoms on almost all the commercial cultivars of apple being grown in Himachal Pradesh. However, symptoms are generally more pronounced on Golden Delicious cultivars. Major source of the transmission of the Apple mosaic and Apple chlorotic leaf spot is through the use of scion wood taken from the infected plants for raising nursery plants (Verma, and Bhardwaj, 2003).The premature leaf fall is another new disease that is gaining importance in Uttarakhand. However, in Himachal Pradesh and Jammu & Kashmir, premature leaf fall has established itself as serious disease. Besides these, San Jose scale and woolly apple aphid continue to be serious problem in all the Apple growing areas of Indian Himalayas and still considered as economic pest. Seedling blight, crown gall and hairy root remain the serious diseases and Apple nursery in all the Apple growing states and cause considerable loss.

The incidence of pests is highest in the mid hills and valley areas and lowest in dry temperate regions. In the mid hills, San Jos scale is the most damaging pest followed by European Red Mite, Woolly Apple Aphid and Blossom thrips. In the high hills, Woolly Apple Aphid and European Red Mite are the most damaging followed by San Jose scale, European Red mite and Blossom thrips. In the mid hills and valley areas, high temperature is responsible for high multiplication of San Jose scale and Red Spider Mite owing to more number of generations per year, thus, causing extensive damage to the crop. Indiscriminate use of these pesticides make them less effective and also developed resistance power against them which cause pesticide residues in food that become hazardous for human and animal health (Qadir and Qamar, 2019).

Constraints in IPM Adoption

The constraints to wider adoption of IPM are important considerations in the transfer of IPM technologies. The technologies developed have yet not been exploited at the level to reach the small and marginal farmers of the country to make IPM as an effective alternative to use chemical pesticides. Even though successful non chemical methods for control of crop pests and diseases have been developed, the transfer of this technologies to the farmers and extension workers has been rather slow. The coverage under IPM at present is estimated to be less than 1% of the total of 143 million hectares of cultivable area in the country. The extent of adoption of the recommended IPM technologies by the farmers depends upon various factors as well as constraints faced by them. Constraints refer to the items of difficulties faced by the farmers in actual adoption of recommended IPM technologies causing technological gap (Luckman and Metcalf, 1994). Constraints are obstacles that farmer experience when using recommended IPM methods, resulting in a technology gap. This review is entirely based on efficient pest control tactis in apple pest control. The control tactics could be used in IMP programmes to supplement PPP-based pest. The Soil characteristics and health can reduced the occurrence of pests within a crop and varies between farm and location. Before establishing of an orchard, it is beneficial to identify the soil type and its properties to give an indication of qualities such as drainage, soil composition and soil quality. These factors may impact fruit production after orchard establishment and determine the most appropriate pre-planting soil amendments (Shaw *et al.*, 2021). The study reveals that high level of complexity of the decision processes make IPM requires careful evaluations, both economic and environmental, considering benefits and costs associated with a good management action.

Some technologies play an important role in IPM is as follows:

- Ignorance of improve cultural practices in the orchard
- Lack of basic knowledge about insect-pests and their time of appearance among fruit growers
- Adoption of improper plant protection measures
- Ignorance as well as poor transfer of technology to farmers on integrated approach of pest and disease management
- Availability of host-plants all the year round
- Indiscriminate use of pesticides
- Inadequate spray technology
- Poor information on pest and disease monitoring among orchardists
- Lack of effective biocontrol agents
- Poor water and nutrient management
- Adulteration of pesticides
- Lack of well organized, systematic and scientific surveillance system to give advance warning to the farmers
- Lack of systematic efforts to solve the problems through multidisciplinary approach

W

Nursery is one of the important components of Apple production. There is a lot of demand for quality, genuine and pest & diseases free planting material. Nursery is a place where young crop plants are raised under intensive care for later transplanting to the field. The use of quality nursery plants is the pre-requisite for the success of any orcharding enterprise. Therefore, the production of healthy and pests-disease free planting material of high quality deserves special attention. The number of genuine and quality Apple plants varieties produced by Agricultural Universities, State Departments of Horticulture, Research Institutions and Private Sectors are very meager to meet the demand of large number of growers throughout the country. For this reason, most of the Apple growers, orchardists are depending on private nurseries and nurserymen for their requirement of Apple. Hence, there is a greater scope to prosper for the nurserymen, who can supply genuine and high quality pests and diseases free planting material of different varieties of Apple plants produced on sound scientific lines.

Such high quality planting material could be produced by careful planning, establishment and management of nurseries. The success of any nursery depends largely upon several factors viz., proper selection of site, careful planning, selection of varieties which are in good demand and procuring the progeny from the reputed institutions/ nurseries, elite gardens, establishment of mother plants block/ scion banks, acquiring scientific information of each variety and skillful management in production of necessary scions as well as marketing the same.

Following pest and disease management practices should be adopted at nursery stage or during establishment of Apple nursery.

- Raise nursery only in solarised soil. In hilly regions, the area which has been identified for nursery raising should be moistened and covered with polythene sheet during May up to sowing period which is done in November. It is also important to cover all 4 sides of polythene with soil or any other material in order to make it air tight and also to increase the temperature of covered area. This helps in controlling all kinds of soil borne pathogens and pests.
- Before four weeks of sowing of seeds in nursery during November, apply *Trichoderma viridae* @ 1 g per kg of FYM and *Beauveria bassiana* @ 1 g per kg of FYM separately in same day. The powder of both the antagonists should be thoroughly mixed in FYM separately. This technique is very effective in controlling soil borne diseases and pests in nursery. Besides, FYM also useful for providing nutrition to the plants for ideal vegetative growth.
- It has been observed that infested nursery plants are main source for spread of the woolly aphid and San Jose scale; therefore, soil should be thoroughly treated with fipronil @ 5 g/m^2 before raising the nursery. Healthy plants should be selected from nursery and is treated with imidacloprid 0.5 ml/litre of water before planting in the orchard.
- Approach grafting has been recommended for the control of collar rot. Raising the grafting point at least 20 cm above the ground level and improved drainage helps in reducing collar rot (Verma and Bhardwaj, 2003).

- For the management of seedling blight in nursery, apply *Trichoderma viridae* or *T. harzianum* @ 0.4% in the soil or drench the soil with thiram 0.3% or aureofungin 0.04%.
- Crown gall disease can be managed by molecular characterization of agrocin and non-agrocin producing strains of nonpathogenic *A. tamefaciens* strain K-84 biocontrol agent.
- Apply *Pseudomonas fluorescence* at 0.4% in the soil for management of hairy root disease in nursery.
- Do not grow the nursery at the same site every year.
- Destroy the affected seedling

Rootstock of Apple

Rootstock are those plant genotypes that make use of their inherent characters like dwarfness, earliness, fruit quality improvement, biotic and abiotic stress tolerance, etc. transmit on the suitable scion variety rootstock. They are divided in two groups as seedling rootstock that based on propagation method and it develop from germinated seeds and other is clonal rootstocks that are vegetatively propagated in nature. *Paron* (*Malus baccata* var. *Himalaica*) should be used as a rootstock for proper management of white root rot. It is also very helpful for eco-friendly disease management system and also effective in reducing the use of chemical pesticides. Fungicides are generally used in India for management of white root rot (Kumar *et al.* 2015, Kumar *et al.* 2014, Rana *et al.* 2010). According to Rayees *et al.*, 2019, rootstocks and propagation is the main backbone of apple cultivation. Different vigour controlling rootstocks along with resistance to biotic and abiotic stress are available.

Table 1: Potential rootstocks of apple resistant against biotic and abiotic stresses

Type of Nature	Rootstock	Promising Characteristics
Disease Resistant	*Malus bacatta var himalica* (Maxim.) Schneid	Partially resistant to collar rot, root rot and wooly aphid.
	Malus bacatta (Linn.) Borkh Shilong	It shows high degree of resistance to powdery mildew.
	M7	Resistance to fireblight and crown rot.
	M116	Resistant to *Phytophora* (the fungus which causes brown rot / collar rot) and woolly aphid and also resistant against re-plant disease.
	P 1, P2, P 16, P-18	All these rootstocks are resistant *to powdery mildew disease.*
	B-9, B-491, MAC-9, O-3, P-2, No vole, G-30, G-65.	All these rootstocks are Resistant to crown and root rot
	B-9, MAC-9, C-6, No vole	All these rootstocks are *Resistant to latent viruses*

Type of Nature	Rootstock	Promising Characteristics
Cold Hardy	M9	It is a cold hardy and dwarf root stock that produces trees approximately 45 percent of their regular height at full maturity.
	M 26	One of the most cold-hardy of the Malling-series rootstocks and also produces a good productive tree.
	B-491, B-490, B-9, O-3, P-2, P-18, P-22, K-14, Novole, Alnarp 2, Robusta 5	All these rootstocks are Cold hardy rootstocks.
Dwarfing rootstock	M27	It is a dwarf rootstock that reaches its mature size within 2-3 years.
	Malus bacatta (Linn.) Borkh (Giabung)	It shows more dwarfing with Golden delicious scion than M9.
	Malus bacatta (Linn.) Borkh (Rohru)	It is from Rohru H.P and an important dwarfing rootstock.
	B-490, B-491, J-9, M-7, M-9, M-26, M-27, MAC-1, MAC-9, MAC-39, MM-106, O-3, O-8, OAR-1, P2, P18	All these rootstocks are *Dwarfing rootstocks*
High Soil pH Tolerant	M-9, MM-106.	Both rootstocks are *tolerant to High soil pH.*
Climatic Resistant	CITH-AR	This rootstock has better suitability to climatic conditions of the North Western Himalayan regions of Kashmir

Production of Virus-free Nursery Plants

Viruses are known to cause massive losses ranging from 30-50 per cent in apple crop all over the world, reducing productive age of the plant in addition to yield loss. The importance of these diseases is increased many folds when the virus goes untested due to their latent nature. Many of apple viruses like apple mosaic, apple chlorotic leaf spot, apple dapple etc. are not vector transmissible and most can be continued by clean nursery management practices.

The presence of viruses in the mother stock can be tested by highly sophisticated techniques like Electron microscopy, Enzyme Linked Immunosorbant Assay (ELISA) or by Biological indexing using indicator hosts. The procedures for maintenance of virus tree nursery stock are given below :

- Use only virus indexing propagating material
- Establish and maintain virus tested mother stock.
- Determine the presence of virus vector in or near nursery area and apply control measures, if necessary.
- Avoid re-grafting of rootstock where first bud take has failed.
- Avoid top working the mother plants.

- Maintain recommended isolation and adjust free spacing to minimize the possibility of natural root grafting.
- Remove wild hosts growing in or near virus indexed plantings.
- Use only sterilized propagation tools.

Pest and Disease Management Practices During Establishment of Orchard

Integrated Pest Management requires through pre-planting even before establishing the apple orchard which includes selection of suitable site, soil type, regular pest and weather monitoring in the orchard, balanced use of fertilizers, water management, proper training and pruning, improved cultural practices, augmentation and conservation of parasitoids and predators, application of biopesticides, neem based formulations and need based application of selective pesticides.

Deep fertile, well drained soils rich in organic matter and having pH 6.5 to 6.8 are mot suitable for apple. There should not be big rocks or impervious calcareous soil plan restricting the root growth. The sub soil is more important than the surface soil from the point of view of orcharding as it is the major nutrient and moisture support to the plants. For apple loam to clay loam soils, rich in organic matter with high water retention capacity are more suitable.

However, the hilly areas where the apples are suitably grown, these systems can not be followed easily. For such sloppy lands the contour system or terrace planting is most suitable and convenient. At the time of planting, the planting distance according to tree vigor, rootstock and soil fertility is to be decided. The placement of pollinizers wherever required are also to be planned at this stage. The only time of layout, digging of pits and planting of apple plants is winters.

In hill slopes where erosion and run off are a serious problem, it is advisable to plant on the contour. On the same elevation the trees are usually planted at the same row in contour. The procedure helps to build up a low terrace at the tree row which minimizes the water run off.

- For plantation of apple in the orchard, digging of pit should be done of 1 x 1 x 1 m size in the month of September-October and plantation should be made in January. Before one or two days planting, the pit must be thoroughly treated with fipronil @ 10 g per pit. The soil which is to be filled in the pit should also be treated by mixing fipronil 20 g in the soil which is needed for one pit.
- Before plantation, the nursery plants must be treated with imidacloprid solution at 0.5 ml/litre water for 20 minute to check the spread of woolly aphid, San Jose scale and other soil borne pests.
- Plants which are infected with root rot, crown gall, hairy root or any other soil borne diseases should not be planted in the orchard.
- Before plantation in the orchard, all the nursery plants must be dipped in *Trichoderma viridae* or *T. harzianum* solution(10 gm/litre water) for one hour. This treatment takes care of soil borne diseases like crown gall, hairy root and root rot.

Pest and Disease Management Practices in Established Orchard

- Orchard sanitation should be given top priority. Sanitation is a simple, but important preventive measure of pest and disease control. Even simple removal of alternate hosts, weed hosts and crop residues greatly reduce the infesting potential.
- During pruning of apple trees in December- January, also prune and destroy the diseased and infested twigs and highly affected thin branches. Use Chaubatia Paste or Copper oxychloride 0.3% after pruning on the cutting portion. Correct pruning allows adequate air circulation round the tree canopy, thereby modifying the microclimate and reducing pest and disease development
- Digging the basins of the tree with spades during winter (October-November) to expose the diapausing pupae and spores to adverse climatic conditions.
- Over wintering caterpillars should be collected from the loose bark, other debris, bands used for trapping and destroyed.
- Apply lime sulphur @ 1:20 (one litre lime sulphur in 20 litre of water) every year in December or January after pruning. It helps in managing powdery mildew, Black spots, sooty blotch and stem diseases.
- Mulching with dry grass or hay (10-15 cm in thickness) in the tree basis during March should be done to conserve the moisture during fruit setting and fruit development period.
- Two to three bee colonies per ha should be provided in each orchard at the time of bloom to increase pollination.
- Fallen fruits should be collected throughout the season and either buried deep in the soil or destroyed otherwise.
- If the apple trees are infested with San Jose scale or woolly aphid, mark the infested trees and spray them with thiamethoxam 0.05%. This application also takes care from woolly aphid. At the same time, spraying the roots and exposing the tree basin with thiamethoxam 0.05% or imidacloprid 0.007%. These applications should be done after pruning in January or February. The root treatment is also very useful in controlling root borer and collar zone weevil (Singh, and Lal, 2002. Singh, *et.al.* 2001).
- All loose bark of apple trees should be scraped off to remove over wintering sites for the caterpillars. Vicinity of trees should also be kept clean of packing cases and all other debris which are likely to shelter the over wintering caterpillars of codling moth during August to mid October.
- In areas where natural population of predatory Coccinellids is high during summer (April-June), application of toxic chemicals should as far as possible, be avoided.
- In case of tent caterpillar infestation during April to June, mopping up the 'tents' with a pole and some rags dipped in kerosene tied on its end is

also helpful. Kerosinized water in an open vessel should be placed below the tree so that caterpillars fallen may also be killed readily. During severe infestation, nimbicidine 0.05% or *Bacillus thuringiensis* (Bt) based bioinsecticides @ 0.2 % can be sprayed (Singh and Pandey, 2004).

- For management of premature leaf fall, spray dodecylguanidinium acetate 0.1% during fruit set (pea size), second with mancozeb 0.2% during fruit development (walnut size), third with carbendazim 0.1%, after 20 days of second spray and fourth spray with mancozeb 0.2% at 20-25 days before picking of the fruits. If level of disease incidence is low, number of sprays can be reduced. These sprays also take care of apple scab and fruit spots (sooty blotch and fly speck). Therefore, this spray schedule must be followed to manage these three important diseases which are regular in occurrence in all the apple growing regions of the country· (Verma and Bhardwaj, 2003)
- Collection and destruction of fallen leaves from the orchard floor during winter are found effective against premature leaf fall disease.
- Apply white wash impregnated with glue regularly on the trees trunks to avoid sun burn and winter injury.
- Spraying with Urea 5% on the leaf litter also helpful in reducing the primary inoculums of premature leaf fall disease by enhancing decomposition.
- If leaves are infected with powdery mildew, spray the trees with dinocap 0.05% or thiophanate methyl at 0.1% or hexaconazole @ 0.1 % .Spraying can be repeated, according to need (Pedigo, 1991).
- Collect and destroy grubs of root borer during October-November coinciding with the preparation of tree basins.
- For management of root rot disease, the basin of the infected tree is brought to a moisture level of 30-40 per cent and 15-20 cm deep holes at a distance of 30 cm are dug with a cow bar all over the basin. These holes are closed after filling with 0.1% suspension of carbendazim. The treatment must be repeated 3 to 4 times every year during the rainy season.
- Application of *Trichoderma viridae* @ 40 g per tree has shown good result against white root rot and collar rot at initial stage.
- In case of collar rot, painting of wounds with copper paints and drenching the area around the trees with mancozeb or metalaxyl + mancozeb 0.3%. Infected trees should be marked and treated in October.
- Painting of wounds with copper paints and drenching the area around the trees with Mancozeb or Metalaxyl plus Mancozeb 0.3 per cent has proved to be quite effective against collar rot.
- Correct training and pruning at the right time of year helps in reducing apple cankers considerably. In south or southwestern exposed sites, the trunk height must be kept short by allowing the first scaffold branches to grow at thirty cm above the ground, to protect the main trunk from

direct sun. Heading back branches of more than 2.5 cm diameter should be done as close to the main trunk as possible without creating a stub. The cut should be clean, slanting and sharp. All the pruning cuts should be covered with Chaubatia Paste or Bordeaux or Blitox-50 WP.

- Tree badly infected with canker disease should be sprayed after fruit harvest with 0.3 per cent copper oxychloride or 0.3 per cent captan. To treat existing canker wounds, first scarify, then wash with five per cent formaldehyde or rectified spirit and apply cow dung paint or Chaubatia or Bordeaux paint afterwards.
- For the control of stem borer, clear the holes with flexible wire and then insert 0.5 gm of paradichlorbenzene (PDCB) in the hole and plug it with puddling material or insert cotton wick soaked with petrol 1 ml per litre of water or dichlorovos 1.5 ml per litre water and plug the hole.

Table 2: Economics of Apple Orchard under High Density Plantation at Billekh village of Almora District of Uttarakhand

Variety	Yield (q/ha)	No. of Plants /ha and Planting distance	Average Sale Price (Rs/q)	Gross Cost(Rs in lacs/ha)	Gross Income (Rs/ha)	Net Income (Rs./ha)	B:C Ratio
Super Chief	666.67	3333(3x1m)	7000	14.00	46.66	32.66	3.33
Red chief	733.26	3333(3x1m)	7250	17.01	53.16	36.15	3.12
Jeromine	699.93	3333(3x1m)	7850	18.13	54.94	36.81	3.03
King Rot	800.00	3333(3x1m)	7100	19.80	56.80	37.00	2.86
Red Velox	733.26	3333(3x1m)	8250	20.56	60.49	39.93	3.00
Scarlet-2	732.32	3333(3x1m)	7600	17.80	55.65	37.85	3.12
Golden Delicious	689.76	3333(3x1m)	8150	18.54	56.21	37.67	3.03
Granny Smith	823.23	3333(3x1m)	7300	18.62	60.09	41.47	3.23
Dark Barrongala	799.12	3333(3x1m)	7050	18.58	56.33	37.75	3.03
Red Lumgala	766.59	3333(3x1m)	7800	20.00	59.79	40.00	3.00
Red Gold	833.25	3333(3x1m)	7250	20.00	60.41	40.41	3.02
Wance Delicious	662.54	3333(3x1m)	7980	17.44	52.87	35.43	3.03
Mitch Gala	739.92	3333(3x1m)	7400	18.61	54.75	36.14	3.00
Gayle Gala	740.21	3333(3x1m)	7660	18.14	56.70	38.56	3.13

Table 3: Major pests of apple in Indian Himalayas

Common Name	Scientific Name	Order	Family	Plant part damaged	Period of activity	Distribution
San Jose scale	*Quadraspidiotus perniciosus* Comstock	Hemiptera	Diaspididae	Twigs, Branches, Fruits	March-December	All apple growing states
Woolly Apple Aphid	*Eriosoma lanigerum* Hausmann	Hemiptera	Aphididae	Twigs, Branches, Roots	Throughout the year	All apple growing states
European Red Mite	*Panonychus ulmi* Koch	Acari	Eriophyidae	Leaves	May-September	Himachal Pradesh
Blossom Thrips	*Thrips flavus* Schrank	Thysanoptera	Thripidae	Flowers	February to April	All apple growing states
Codling Moth	*Cydia pomonella* Linnaeus	Lepidoptera	Torticidae	Fruits	April to September	Apple areas of Ladakh
Apple Root Borer	*Dorysthenes hugelii* Redtenbacher	Coleoptera	Cerambycidae	Roots	Throughout the year	All apple growing states
Apple stem borer	*Apriona cinearea* Cheverlot	Coleoptera	Cerambycidae	Stems	March to October	All Apple growing states
Collar zone weevil	*Dyscerus fletcheri* Marshall	Coleoptera	Curculionidae	Collar zone region of the tree	March-October	Uttaranchal
Tent caterpillar	*Malacosoma indica* Walker	Coleoptera	Lasiocampidae	Leaves	March-May	All apple growing states
Fruit Moth	*Argyresthia conjugella* Zeller	Lepidoptera	Yponomeutidea	Fruits	June-August	Himachal Pradesh
Indian gypsy moth	Lymantria obfuscata Walker	Lepidoptera	Lymantridae	Leaves	March-May	All apple growing regions

The data pertaining to economics of Apple cultivation under High Density Plantation of commercially important varieties having demand in domestic and International market was recorded at Billekh village of Almora district of Uttarakhand in seven year old orchard. The Plantation was done at 3x1m, thus total no of plants in a hectare were 3333 including 30 per cent pollinating varieties such as Golden Delicious, Red Gold , Granny Smith, Red Lumgala etc. The produce was sold in the Delhi markets and received premium price ranging from Rs. 7000 to 8250 per quintal. The gross cost varied from Rs. 14.00 to 20.56 lacs per hectare. The gross income was also calculated and found that it was between Rs. 46.66 to 60.49 per hectare. According to an assessment maximum Net Income received from Granny Smith variety i.e. Rs. 41.47 lacs per hectare and minimum was in Super Chief which was Rs. 32.66 lacs per hectare. Accordingly B: C ratio ranges from 2.86 to 3.33.

Table 4: Major diseases of apple in Indian Himalayas

Common Name	Scientific Name	Plant Part damaged	Period of Activity	Distribution
Premature leaf fall	*Marssonina coronaria* Syn.	Leaves	July to August	All apple growing states
White Root Rot	*Dematophora* necatrix	Roots	Round the year	All apple growing states
Collar Rot	*Phytophthora* cactorum	Collar region of the tree	Throughout the year	All apple growing states
Apple Scab	*Venturia inaequalis* Cke.	Leaves and Fruits	April-August	All apple growing regions
Apple Mosaic	*Virus*-Normal	Branches, Leaves	July-September	All apple growing states
Canker Diseases	*Botryosphaeria mali, Coniothecium chomatosporum, Corticium salmonicolar*	Trunk, Twigs, Branches	Round the year	All apple growing states
Fruit Spots	*Gloedes pomigena*	Fruits	June to August	All apple growing states
Powdery mildew	*Podosphaera* leucotricha	Shoots Buds, Leaves	April to August	All apple growing states
Apple Chlorotic leaf spot	*Virus* Normal	Leaves, Frutis	July-September	Himachal Pradesh
Apple Dapple	*Viroid Pathogen-* Normal	Fruits	July-September	Himachal Pradesh
Seedling blight	*Sclerotium rolfsii* Sacc.	Roots	April-August	All apple growing regions
Crown Gall	*Agrobacterium tamefaciens* Smith and Townsend	Roots	August-September	All apple growing regions
Hairy Root	*Agrobacterium rhizogenes* Ricker and Ricker	Roots	August-September	All apple growing areas

Table 5: Effective Natural Enemies of Major Pests of Apple Crop in India

Pest	Natural Enemy	Parasitoid/ Predator	Method of Application
San Jose scale (Quadraspidiotus perniciosus Comstock)	*Encarsia perniciosi* Tower	Parasitoid	2000 adults per infested tree
	Aphytis sp. (Proclia group)	Parasitoid	50 adults/ grubs per infested tree
	Chilocorus bijugus	Predator	50 adults/ grubs per infested tree
Woolly Apple Aphid (Eriosoma lanigerum Hausmann)	*Aphelinus mali*	Parasitoid	1000 adults per infested tree
	Chrysoperla carnea	Predator	50 adults/ grubs per infested tree
	Chrysopa scelests	Predator	50 adults/ grubs per infested tree
	Coccinella septumpunctata	Predator	30-50 grubs/ adults per infested tree
Codling Moth (Cydia pomonella Linn.)	*Trichogramma embryophagum*	Parasitoid	2000 adults per infested tree at the interval of one week
European Red Mite (Panonychus ulmi)	*Chrysoperla carnea*	Predator	50 adults/grubs per infested tree
Thrips (Thrips flavus)	*Anthocoris minae*	Predator	30-50 grubs per infested tree

Table 6: Biocontrol agents antagonists to plant pathogens attacking apple crop in India

Apple	Antagonist	Plant parts where antagonist applied	Pathogen suppressed
Grey mould	*T. harzianum*	Flowers	*Botryis cinerea*
Grey mould	*T. pseudokoningli*	Flowers	*Botryis cinerea*
Crown gall	*Agrobacterium radiobacter* Strain K-84	Seed and root	*Agrobacterium tumefaciens*
Scab	*Athelia bonbacina*	Fallen leaves	*Venturia inaequalis*
Collar-rot	*Enterobacter aerogenes, Bacillus subtilis*	Root	*Phytophthora coctorum*
Collar-rot	*Trichoderma* spp. and *Gliocladium*spp.	Root	*Phytophthora coctorum*
Cankers	*Bacillus subtilis*	Leaf, scar, wounds	*Nectria galligena*
Cankers	*Cladosporium cladosporoides*	Pruning cuts	*Nectria galligena*
White root rot	*Enterobacter aerogenes*	Root	*Dematophora necatrix*
White root rot	*T.viridae and T.harzianum*	Root	*D.necatrix*

Apple	Antagonist	Plant parts where antagonist applied	Pathogen suppressed
Silver leaf	*T.viridae*	Tree wounds	*Chonodrostereum purpureum*
Powdery mildew	*Ampelomyces quisqualis*	Leaves	*Podosphaera leucotricha*
White root-rot	*Glomus spp.*	Root	*D.necatrix*

Challenges for the Future

A review of information available on various aspects of management of insect pests and diseases of apple indicate that there is tremendous potential for the adoption of IPM schedules against economically important pests and diseases. However, no attempt has been made so far to integrate and recommend the same so as to make it practically viable in the orchard (Kogan, 1998). Among the various components of IPM, cultural practices are one of the most sound and practicable eco-friendly approaches. It includes selection of appropriate variety on suitable rootstock, time and manner of planting, tillage, orchard sanitation, application of irrigation and fertilizers, proper training and pruning, harvesting time and procedures and even off-season operations in the orchard. These practices need to be modified and refined to make them compatible with intensive horticulture to maximize the quality and productivity of apple. A number of potential natural enemies have been identified that could be effectively utilized for the management of major pests, especially San Jose scale, Woolly aphid and peach leaf curl aphid. However, methods are to be standardized for the mass multiplication of potential natural enemies and their evaluation in the orchard. There is a need to establish a network of large scale natural enemy multiplication units, so that these are made easily available to the farmers. Very few microbial agents have been identified against major insect-pests and diseases of apple. Screening and evaluation of microbial pathogens are to be taken up on a priority basis. The high humid condition under apple plants are grown is a positive factor for the potential use of microbial agents. Chemical control involving use of insecticides continues to be the main component of IPM in apple but their application could be considerably reduced to need based application of selective insecticides based on surveillance of pests, improvement in formulation and application technology. There is an urgent need to determine appropriate ETLs for specific crop pests situations in different agroclimatic region of the country. Farmers should also be educated on ETL and taught to use it for deciding their spray application. Pheromone traps and lures have been developed for a number of insect-pests. These could be used for monitoring adult activity and egg laying. Very little information is available on the effect of various plant extracts on major insect pests. Neem and other plant extracts are to be evaluated and if found promising could be used as component in IPM schedules. In the end, it may be stated that there is a wide gap in available technology and that being adopted by farmers. This gap can be minimized and available technology implemented in true letter and sprit by convincing the farmers about benefits of IPM programmes, its need and disadvantages of indiscriminate use of pesticides. Innovations in farmers participation and proper training can help to overcome many of the implementation problems. Small-scale farms are an important component of agricultural production even in developed economies, social development with conservation of agricultural biodiversity and fulfill the demand of food security. This review details some of the characteristics of the small farm sectors in developed economies and identifies some of the characteristics of small farms and their operators that may favor the implementation of IPM. Knowledge gaps and opportunities are responsible for IPM outreach to small-scale producers. (Grasswitz, 2019)

Future Prospects

From the last one and half decades considerable progress has been made in the development and promotion of IPM in apple at national level under the various projects and policies. There is quite a lot of information available about IPM in Indian Himalayas where Apples are grown and wherever IPM practices adopted in true letter and spirit, farmers harvested good quality Apples with minimum use of pesticides. It is also true that IPM packages developed are mainly based on combination of improved cultural practices, biopesticides and need based application of safer and selective pesticides.

The studies conducted under the World Bank funded project of Diversified Agriculture Support Project show that most of the IPM practices including pesticides in apple are done during dormant stage (October-March) of the tree and if these practices are adopted by the farmers, the incidence of pests and diseases during flowering, fruit setting and fruit developmental period of apple (April-August) is very less for which available biopesticides or 2-3 applications of appropriate pesticides on need basis are quite enough to manage pests and diseases during these crucial stages.

Sometimes it is found that only biopesticides are sufficient to keep them below the economic damage level. Thus, implementation of given IPM strategies does not cause any adverse impact on non target organisms and undesirable residue in the fruits. Therefore, IPM practices which have been mentioned in the manuscript show good perspective in apple.

Unless farmers acquire basic knowledge about identification of pests and diseases, their time of appearance, peak period of incidence, nature and extent of damage and suggested/ recommended IPM practices, it is impossible for them to resort to apply appropriate/ relevant practices for effective and integrated management of various pests and diseases. The discussion held with farmers on above aspects during study indicated that most of the farmers are still not aware about the use and importance of recommended IPM practices hence their implementation in Apple orchards is poor despite considerable progress in IPM and its good performance in existing Apple ecosystem in the country.

With a view to promote greater IPM implementation, special emphasis is necessary to generate increased awareness and transfer of recommended IPM practices. The strategic steps will include determination of farmers' pest management knowledge, attitude and practices, IPM trial demonstrations and appropriate training of extension personnel and farmers.

The observations recorded in the Apple orchard from different varieties revealed that under High Density Plantation of above varieties in the prevailing Agroclimatic conditions fetched remunerative return even for small and marginal farmers at the altitude of 1700-2000 above MSL. The studies carried out in the Apple orchard at Billekh village of Almora district of Uttarakhands indicated that implementation of various practices of Integrated Crop Management including Pest and Disease management can provide profitable and sustainable income to the farmers of Himalayan region of India. Keeping in view the present existing scenario and growing demand of residue free Apple fruits in the national and international market, encouraging policies and programmes of Central and State Governments and their commitment for promotion and advancement of IPM reveals its bright perspective in the country.

References

Altaf S, Ahad I, Pathania SS, Lone GM, Peer FA, Maqbool S. (2019) Insect pest complex of apple nurseries in North Kashmir. *Journal of Entomology and Zoology Studies* Anow (2019): 7(3):697-700.

APEDA Agri Exchange (2021). Database https://agriexchange.apeda.gov.in.

Apple Production Method and Process Annals of R.S.C.B., ISSN:1583-6258, Vol. 23, Issue 2, 2019, pp. 118 – 127

Bhalla, O.P. and Gupta, P.R. (1993). Insect-pests of temperate fruits. *In* : K.L. Chadha and O.P. Pareck (eds.)., *Advances in Horticulture – Fruit Crops*, 1993, Vol. 3, Malhotra Publishing House, New Delhi, pp. 1557-1589.

Coll M, Wajnberg E (2017) Environmental Pest Management: A Call to shift from a Pest-centric to a system-Centric Approach. Environmental Pest Management Challenges for Agronomists, Ecologists, Economists and Policymakers, 1-17.

Deguine, J. P., Aubertot, J. N., Flor, R. J., Lescourret, F., Wyckhuys, K. A., & Ratnadass, A. (2021). Integrated pest management: good intentions, hard realities. A review. *Agronomy for Sustainable Development*, *41*(3), 1-35.

FAO (2020) NSP – Integrated Pest Management, FAO definition.

http://www.fao.org/agriculture/crops/thematic-sitemap/theme/pests/ipm/en/.

Gaikwad M. B., Challa N., Panma Y. and Priyanka T. (2019). Push-pull strategy: Novel approach of pest management *Journal of Entomology and Zoology Studies*; 7(5): 220-223.

Georgiadis N, Tsarouhas K, Tsitsimpikou C,Vardavas A, Rezaee R, Germanakis I, TsatsakisA, Stagos D and Kouretas D (2018). Pesticides andcardiotoxicity. Where do we stand? *Toxicol ApplPharmacol*. **6**: 1-14.

Grasswitz, T. R. (2019). Integrated pest management (IPM) for small-scale farms in developed economies: Challenges and opportunities. *Insects*, *10*(6), 179.

Gupta, P.R. (1999). Rerospect and Prospect of Biological Control of insect-pests of temperate fruits in India. In : Proc. National Seminar on Himalyan Horticulture. Jagmohan Singh, R.P. Sharma and V.K. Sharma (eds.) *Hortivision*, 1999, pp. 298-307.

Hussain, B., Buhroo, A. A., War, A. R., & Sheerwani, A. (2018). Insect-Pest Complex and Integrated Pest Management on Apple in Jammu and Kashmir, India. In: Apple: Production and Value Chain Analysis. Astral: ISBN: 9789351249214.

Kakar, K.L. (2000). Pest status of temperate fruits and their management. *In* : P.R. Gupta and K.K. Jindal (eds.) *Horticulture Technology* : Vision -2000 and Beyond, 2000, 2 pp. 166-178.

Kogan, M. (1998). Integrated pest management : Historical perspectives and comtemporary developments. *A. Rev. Ent.*, 1998, 43 : 243-270.

Kumar A, Attri B L, Kishor A, Narayan R, Debnath S and Ahmed N. (2015). Integrated management of white root rot of apple caused by *Dematophora necatrix* Hartig. (*In*) *National Conference of Temperate Fruits and Nuts- A Way Forward for Enhancing Productivity and Quality*. The Horticulture Society of India, New Delhi, p 194.

Kumar A, Attri B L, Raj Narayan, Kishor A, Debnath S, Ahmed N and Mer M S. (2014). Identification and management of *Dematophora necatrix* Hartig causing white root rot of apple. *Souvenir of 6th Indian Horticulture Congress-2014*, 6-7th November, 2014, Coimbatore, Tamil Nadu, pp 395–396.

Kumari D and John S. (2019) Health risk assessment ofpesticide residues in fruits and vegetables fromfarms and markets of Western Indian Himalayanregion. *Chemosphere*. **224**: 162-167.

Luckman, W.H. and Metcalf, R.L. (1994).The pest management concept. In : R.L. Metcalf and W.H. Lucknmann (eds). *Introduction to Insect Pest Management.* John. Wiley & Sons, New York, 1994, pp. 1-34.

Pedigo, L.P. (1991). *Entomology and Pest Management.* MacMillan Publishing Co., New York, USA, 1991.

Qadir I. and Qamar A. (2019). Monitoring of Mite Population and Pest Management Practices in some Apple Orchards in Kashmir Region. *Biosciences Biotechnology Research Asia*, 16(2), 417-424

Rana S, Sharma S K and Sharma J N. (2010). Integration of cultural, chemical and biocontrol methods in management of white root rot of apple. *Indian Phytopathology* **63**(2): 207–211.

Rathee M. (2018) Emerging insect pests in Indian agriculture. *Indian Journal of Entomology;* 80 (2): 267-281.

Rayees Afzal Mir, Syed Aasif Hussain Andrabi, Mohd Gulfishan. (2019) Rossi, V., Sperandio, G., Caffi, T., Simonetto, A., & Gilioli, G. (2019). Critical success factors for the adoption of decision tools in IPM. *Agronomy*, *9*(11), 710.

Sharma, J.N. (2003). Managing Premature Leaf fall in apple. *In :* K.K. Jindal, Rajan Bawa and Tej Pratap (eds.), *Apple Farming and Livelihoods in the Himalayas.* Bishan Singh Mahendra Pal Singh Press, Dehradun, 2003, pp. 289-292.

Sharma SK. (2018). Apple High Density Orcharding – Performance Reckoning Studies in Respect of the Agroecological Challenges of Kinnaur. *International Journal of Science and Research*;8(4):636-642.

Shaw, B., Nagy, C., & Fountain, M. T. (2021). Organic Control Strategies for Use in IPM of Invertebrate Pests in Apple and Pear Orchards. *Insects*, *12*(12), 1106.

Singh, J., & Patel, N. R. (2017). Shift from a Pest-Centric to a System-Centric Approach. In: Assessment of agroclimatic suitability of apple orchards in Himachal Pradesh under changing climate. *Journal of Agrometeorology*, *19*(2), 110-113.

Singh, S.S. and Lal, H. (2004). Managing insect-pests on apple scientifically. *Indian Hort.*, 2004, 48 (4) : 22-25.

Singh, S.S. and Lal, S.D. (2002). Integrated Pest Management in Temperate Fruits. In : D.Prasad and S.N. Puri (eds.) *Crop Pest and Disease Management* – Challenges for the Millennium. Jyoti Publishers, New Delhi, 2002, pp. 39-50.

Singh, S.S. and Pandey, J.C. (2004). Integrated Pest Management in Apple. In : H. Lal, S.S. Singh and V.P.Singh (eds.). *Apple Souvenir*. Published by Department of Horticulture on Apple Show held at Nainital from 27-28 August, 2004, pp. 46-50.

Singh, S.S. Srivastava, P.M. and Phogat, K.P.S. (2001). Collar-zone weevil – an emerging problem on apple in U.P. hills, *Indian Hort.*, 2001, 46 (1) : 12 and 15.

Verma, K.D. and Bhardwaj, L.N. (2003). Apple Disease- Based Low Productivity in the Himalayas. *In* : K.K. Jindal, Rajan Bawa and Tej Pratap (eds.). *Apple Farming and Livelihoods in the Himalayas*. Bishan Singh Mahendra Pal Singh Press, Dehradun, 2003, pp. 299-306.

Chapter 17

Latest Technologies in Grape Processing

Ajay Kumar Sharma, Ahammed Shabeer TP, Prashant H Nikumbhe and Nishant A Deshmukh

ICAR-National Research Centre for Grapes, Pune-412307 Maharashtra (India)

Grape is very old known fruit crop. It is believed that grapevine is originated in the Caucasian region (Grassi et al., 2006). The cultivation of domesticated grapes dates back four millennia in the eastern Mediterranean and two millennia in Western Europe. Cultivation to produce fruits, consume as fresh or dried, and especially in form of wine, a drink of great economic, cultural and symbolic value (Brothwell and Brothwell 1969 and McGovern 2003). Basically, grape is a temperate region crop but due to its adaptability, now being grown in wider ranges of climatic and soil conditions. Among the various factors, climate is most dominant to determine the success of viticulture in new regions (Parker et al. 2020). With its several thousand varieties, the *Vitis* genus is today characterised by high levels of genetic diversity. According to Lacombe (2012) actual number of vine varieties from *V. vinifera* species in the world is estimated 6 000 numbers. Presently only few cultivars have higher economic values, however, thousands type grape have been described and classified as per their uses as table purpose, preparation of wine, raisins, juice and other specific products(Alleweldt and Possingham 1988). Of the world's 10 000 known grapevine varieties, only 13 varieties cover more than one-third acreage under grapevines and 33 varieties cover 50% of total coverage (OIV, 2017). The varieties planted in many countries are called as "international varieties" e.g. Cabernet Sauvignon, is known as one of the most cultivated varieties in the world which covers 5% of the total world area. While, few varieties are cultivated on sizeable area but in limited countries i.e. Kyoho, mainly grown in China. Among the varieties Kyoho at top followed by Cabernet Sauvignon, Sultanina, Merlot, Tempranillo, etc. Among these varieties, Kyoho is known consumed as fresh, Sultanina (Thompson Seedless) mainly for table and raisin purpose and others are for wine making.

The global vineyard surface area is estimated to be 7.3 mha in the year 2021 (OIV, 2022) with a production of about 80.1 million metric tons. Presently India stands at 7th rank in total grape production. India is known as table grape growing countries and it is at 2nd just after China. According to an estimate an area of 0.162 million ha produced 3.49 million tons grapes during 2021-22 (Anonymous, 2022).The grape growing is mainly concentrated in tropical regions where hot weather conditions prevail. EU countries are known for wine grape production and these grapes are converted into wines. When we look worldwide, mostly grapes are utilized for winemaking followed by table purpose and raisin making. Worldwide about 48% of total grapes are fermented and wines are being prepared and about 36% grapes are dried for raisin making. But in India about 68% of total produce is consumed as fresh grapes and 30% are used for raisin making while only 1-2% are converted into wines. Very negligible quantity is used for juice purpose. A shift in grape consumption behaviour is observed and consumers are shifting from wines to fresh grape consumption.

The grapes are known for its health benefits. Grape berries contain bioactive compounds having health benefits, but accumulation of these compounds is affected by various factors like as variety, maturity level at the time of harvesting, post-harvest handling and storage conditions, environmental factors such as location, light conditions, temperature, nutrition, water, micro-organisms and followed viticulture practices to obtain the grapes (Chen et al., 2020, Colombo et al., 2020, Perestrelo et al., 2012, Yang et al., 2020). According to a study in the US, grapes and its products consumed by adults or children are associated with healthier dietary patterns and improved nutrient intakes. In the non-alcoholic forms most commonly consumed – fresh grapes, raisins and 100% grape juice (Keast et al. 2011 and McGill et al. 2011). It can be stated that the consumption of grape products is associated with a healthier dietary pattern and higher intake of key nutrients by both children and adults (McGill et al 2013).

According to Merriam-Webster dictionary *Innovation*, for its part, can refer to something new or to a change made to an existing product, idea, or field. In the present era, innovative ideas are becoming reality and each and every corner is supporting innovative ideas. In case of grape growing and grape processing many innovative ideas have become reality and the stakeholders are being benefitted in different terms. The advancements involve innovative development for benefit of stakeholders. Some innovative ideas which are capable to make remarkable changes in the sector of grape processing are highlighted in the present article.

Grape Processing

Worldwide grapes are mainly utilized for winemaking. Countries like Spain, Italy, France, Germany, Portugal, etc. are known as old wine world while new wine world comprises countries in North (US and Canada) and South America (Brazil, Chile, Argentina), as well as South Africa, Australia, New Zealand, etc. These characteristics are typical, but by no means a rule. Old World wine tends to have lighter body, lower alcohol, higher acidity, and less fruity flavor with more minerality. New World wine tends toward fuller body, high alcohol, lower acidity, and pronounced fruit flavors. Grape drying is very old practice of grape preservation and in ancient era raisins were used as currency also. Countries like US, China, Iran, Turkey, Australia, India, Brazil, Afghanistan etc. are main raisin producing countries. Other than these two products grape juice, jam, jelly, verjuice, etc. have very low share in the sector of grape processing.

Wines: On the basis of colour wines are divided in three categories like red, white and rose or blushed wines. Based on juice yield and must quality parameters cvs; Shiraz, Cabernet Sauvignon, Merlot, Zinfandel, Malbec, etc. are among red varieties and Sauvignon Blanc, Ugni Blanc, Chenin Blanc, Chardonnay, Riesling, Vermentino and Garganega are among white varieties are found suitable for quality wine production (Karibasappa and Adsule, 2008 and Sharma et al. 2020). In India, winemaking is very new industry and commercial winemaking was started about 40 years back. Wineries are mainly located in Maharashtra and Karnataka states and both states have own wine policy to support the industry. To encourage this industry and establish the fair and expandable trade environment, the GI of Nashik Valley Wine is also granted. Red and rose wines are made from red wine grapes where crushed grapes are fermented by using desired yeast at a temperature of 20-24 C and the wines are separated from pomace after completion of fermentation. In case of white wines, juice of white wine grapes is fermented by using desired yeast at the temperature of 15-17 C and wines are separated from lees when fermentation is completed. These are basic methods of wine making.

Consumption of alcoholic drinks is very popular in India; hence, alcoholic drinks like whisky, rum and beer are very high in demand. Indian population is not aware about health benefits of wines. There is need to educate the consumers about health benefits and quality of wines. Meanwhile, many countries are adopting the policies to discourage alcohol consumption in daily life. Many groups in EU countries are demanding wines with low or without alcohol. Hence, due to new trends in wine consumption and shift of high alcohol to low alcohol wines,many techniques have been tested for the reduction of ethanol content in wines (Bucher et al. 2019).

The visible changes in climatic conditions are leading to expansion of grapes in new regions, even warmer regions for wine grape production and grapes from these regions may have more potential alcohol. Same time there is need to produce wines with reduced sulfur content (or even sulfur free) without compromising on quality. Hence,planned activities initiated in this direction and several alternative technologies such as pulsed electric fields (PEF), high-pressure processing (HPP), power ultrasound (US), ultraviolet irradiation (UV), high-pressure homogenization (HPH), filtration, and low electric current (LEC) have been identified and are in process of comparison. These technologies have been explored for inactivation of microbial activities and maintaining wine quality (Silva and Van Wyk, 2021). Another potential area of advancement is clarification of wines. Many clarifying agents have allergy problems. Hence, new natural, non-allergenic and non-animal fining agents have been developed by the different players, which helping in obtaining high quality wines (Marangon et al. 2019).Use of silver nanoparticles appears to be very attractive/popular in control of microbes in wine. Nanotechnology is a promising area where multiple applications in the field of wines; however, several scientific, technological, and legislative obstacles must be overcome to establish real in situ applications of this technology in the wine industry (Monge and Moreno-Arribas 2016).There is a critical decision to choose wine bottle closures, as the closures are known to influence the wine post-bottling due to their performance. The main functions of closures are to seal and prevent liquid leakage during storage as well as avoid any type microbial contamination and oxidation of wines. In the study it was found that the sensory panel could not detect any difference in aroma and flavor profile of wines where different closures (screw cap bottles, the synthetic and natural cork bottles) were used (Cantu et al. 2022).

Raisins: Grape drying is centuries old practice. Dried grapes are also known as raisins. Raisins are commercially made by drying grapes gathered in various places, and they can be consumed raw, roasted, baked, or brewed. Raisin variants are also determined by the variety of grape used to make them. Raisins come in various sizes and hues, including black, green, yellow, blue, brown, and others. Raisin are made by natural drying (drying in open fields or drying on vine), drying in shade, mechanized drying, etc. Adoption of particular methodology for grape drying is affected by various factors like available resources, climatic conditions, preferred type of raisins by consumers, utilization of products, etc. In India, Australian grape drying methodology has been adopted by minor modifications. The grapes are dried under shade after dipping in the solution of ethyl oleate and potassium carbonate. Raisin are again processed to make it marketable by cleaning/washing, grading, packing, etc. (Sharma et al. 2018). In India, raisin making is mainly concentrated in Sangli and Solapur districts of Maharashtra and adjoining areas of Vijayapura and Bagalkot districts of Karnataka. The raisins are mainly consumed as snacks in the country while in EU countries and other important raisin consuming countries, the raisins are mainly utilized by bakery, confectionary and dairy industry. Nutritionally no difference in raisins of same variety having green or brown colour but there is high economic significance as it affects the return from the market.

The raisin making areas of India face high temperature as well as scarcity of water. In the process of raisin making a lot of water is utilized and same time, the dipping solution is just thrown in the fileds having ethyl oleate and potassium carbonate. It creates pollution in soil and water. If microcracks on berry skin can be created by any other method instead of bunch dipping in solution (ethyl oleate and potassium carbonate), the water can be saved same time we can save our soil or water from pollution also. Novel alternative technologies having debuncher and abrasive to create microcracks on berries, developed and licenced have potential to dry grapes without using chemicals (Pawar et al 2021a and Pawar et al. 2021b). The quality of raisins prepared by traditional method of dipping in solution and by developed technology of abrasion was found at par. In same manner, the combined effect of pulsation and vacuum enhances the drying rate resulted in lower biochemical reactions inside the grape berries (Zhang et al., 2018).The vacuum in the Pulsed vacuum drying process prevents the browning reactions and oxidation deterioration. This process results in obtaining raisins having good colour,texture and quality. The microwave-assisted air drying (MWD) is also explored and found to reduce grape drying duration with acceptable raisin quality (Khiari et al., 2019).

The microbial load in the raisins may be at levels that may not threaten human health.Generally, *Salmonella* and *E. coli* are not detected in the raisins but total count of microbes creates problem in smooth trade of the product as well as influence the choice of consumes. Application of chlorine dioxide was registered to low the microbial status of raisins in comparison to untreated (Sharma et al. 2017). The raisins lose quality in terms of moisture, colour, texture, structure, microbial status, etc. during storage. Coating of fruit by using various type materials is found to improve overall quality including shelf life also. Coating of raisins with 0.25% guar gum with 40 per cent glycerol extended storage life and maintained good colour with higher antioxidant activities (Jankar et al. 2019). Raisin processing is a new area where different type advanced techniques are being adopted. Actually, raisin processing starts after collation of dried grapes. Online cleaning, washing, grading and packing are important components of this sector. Cleaning and grading were done manually, it was replaced by simple machineries where size-based grading was performed. Introduction of laser-based graders is a stride in raisin industry where size and colour-based grading and rejection of foreign materials being performed. But online process has changed the whole scenario where raisins are cleaned for separation of unwanted materials, washed to remove dust particles and oil residues, grading based on size, colour and acceptance of the material and finally material is packed. This process improves overall appearance, cleanliness, free from unwanted materials and grading.

Juice: Grape juice is very small sector as grape juice is mainly fermented to make wines. Many varieties have been developed for this purpose. Among these varieties Manjari Medika is found commercially acceptable and being planted in different grape growing states of India. This is teinturier type where higher anthocyanin content of 4-6 g/kg is recorded. But, for further the expansion in the area,there is need of support from juice industry. In this sector, remarkable changes have been made to develop new and unique products other than quality improvement and storage without compromising on quality parameters.In case of grape juice preservation many techniques are being utilized. But, pulsed electric field (PEF) is found superior over thermal and HHP which have the advantage of mechanical and pre-pressure techniques while maintaining the bioactive components and nutritional value of grape juices (Kersh et al. 2023). Various high value products like probiotic are being explored from Manjari Medika juice and the probiotic drink made from Manjari Medika juice is found acceptable (Dagadkhair 2022). The developed probiotic drink harnesses the nutritional and therapeutic virtues of Manjari Medika grapes and can also assist processors to diversify their product profile towards the "functional" non-dairy beverages. A natural isotonic beverage based on grape juice with antioxidant capacity and interesting sensory propertieswas developed by Bendaali et al. (2022).

Waste to wealth: Juice and wine industries develop different type wastes like juice lees, skin, seed, wine lees, etc. These generated materials can be utilized for making feed, fuel, ethanol, natural colours/dyes, food additives, extracting phenols/anthocyanins, etc. Very little portion of these wastes (by-products) is being utilized worldwide while these wastes have immense values. There are many cosmetic and functional products are available worldwide based on waste material generated by wineries. But still very big portion of these wastes is dumped for landfill and creating soil and water pollution. Wine lees is very potential and utilized in different ways. It can be utilized again for making quality wines. Wine lees collected from second racking of red wines contained functional compounds. Utilization of processed wine lees obtained from second racking of Cabernet Sauvignon wine improved nutraceutical, rheological and sensory properties of ice cream, low sugar ice cream and yoghurt (Sharma and Aglawe, 2022, Pundhir et al. 2018 and Sharma et al. 2015). Zero waste processing of Manjari Medika grapes was found very suitable (Fig. 1). The juice contained higher anthocyanins and phenols and showed higher antioxidant activities. The juice can be utilized for blending to improve quality of other juices also. The left-over pomace processed again and after drying skin and seeds were separated. Technologies have been developed to utilized skin powder in preparation of cookies, breads and other bakery products. Addition of skin powder replaces fine wheat flour up to 15% and improve antioxidant properties, add natural colour, dietary fibres and make attractive high value bakery products (Sharma and Shabeer, 2019). Same time grape seeds can be utilized for grape seed oil. Vitamin E content (especially tocotrienols) of Manjari Medika seed oil (1.15–1.35 g kg 1) is distinctively higher which indicating superior quality oil having nutraceutical values. Triacylglycerols include trilinolein (LLL, 43%), dilinoleoyl-stearylglycerol (LSL, 19%), and dilinoleoyl-palmitoylglycerol (LLP, 11%), which are earlier recognized as natural antioxidants was found in the seed oil of this variety (Khan et al. 2020). As this variety contains higher anthocyanins, the developed technology for its extraction and purification is able to make microencapsulated anthocyanincapsules through spray drying techniques and can be utilized as preventive measures as it possesses many health benefits. Comparison of skin powders of different juice and wine varieties clearly showed the differences for functional properties and addition of different types of skin powders improved the nutraceutical properties of cookies (Sharma et al. 2018).

Innovative ideas are being adopted for utilization of wastes generated by winery and grape juice based industries. Wine leather (a vegan product) has been developed by Vegea a start-up. Grape marc contained skin and stalkis recycled and dried. Grape seed oil obtained and rest is polymerized by patented methods. By adopting innovative process and adding chemical substances created a material which is very similar to animal leather. Ten liters of wine generates about 2.5 kg of "waste" from which 1 square meter of wine leather can be produced (Kohlbacher 2021). Residue of Pinot Noir grapes placed in microwave and then freeze-dried. When it was turned into powder and added to the molecular matrix of polypropylene, a plastic frequently used for packaging. Untreated plastics started to crack after 25 hours while plastics infused with the grape material lasted twice as long (Diouf-Lewis et al. 2018). Israeli start-up Nutrilees that has fine-tuned a process to remove the unpleasant taste and smell compounds from the wine lees. This startup produced a tasteless, nutrient-packed purple powder. It is high in protein, dietary fibre, polyphenols, and minerals. The powder is deep purple which adds a purple hue to products such as cereals, protein bars, pasta and ice cream(Michail, 2021).

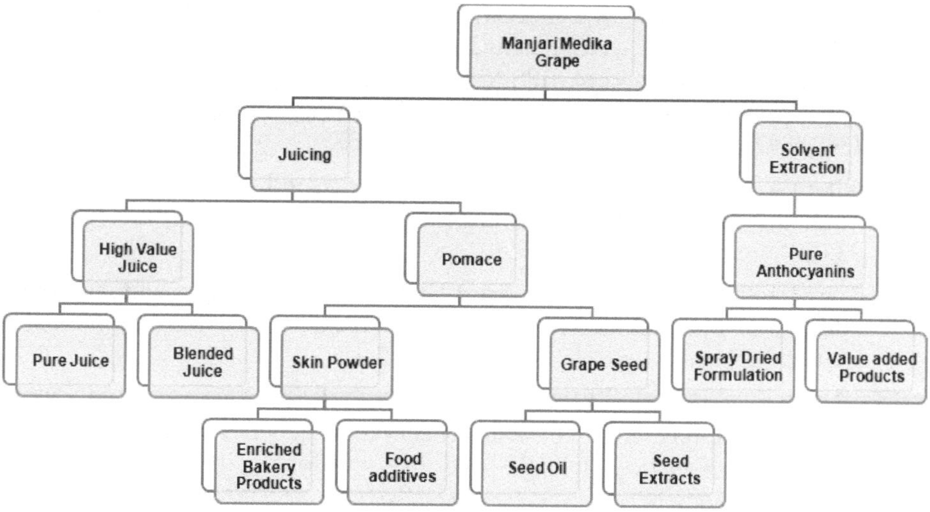

Fig. 1: Zero waste processing of Manjari Medika grapes

References

Alleweldt, G. and Possingham, J.V. (1988). Progress in grapevine breeding. *Theor Appl Genet*,**75**: 669–673.

Anonymous. (2022). 2020-21 (Final) of area and production of horticulture crops category-wise: All India. 2019-2020. Available from: https://agricoop.nic.in/sites/default/files/2020-21%20(Final)%20Advance%20Estimates%20 2020-21%20(1).pdf

Bendaali, Y.; Vaquero, C.; González, C. and Morata, A. (2022). Contribution of grape juice to develop new isotonic drinks with antioxidant capacity and interesting sensory properties. *Frontiers in Nutrition*, **9**: 1-9. Article 890640 https://www.frontiersin.org/articles/10.3389/fnut.2022.890640

Brothwell, D. and Brothwell, P. (1969). Food in antiquity. a survey of the diet of early peoples. London: Thames and Hudson. 248 p.

Bucher, T.;Deroover, K. and Stockley, C. (2019). Production, and marketing of low-alcohol wine. In: Morata A, Loira I, editors. Advances in Grape and Wine Biotechnology. London, UK: IntechOpen; DOI: 10.5772/intechopen.87025

Cantu, A.; Guernsey, J.;Anderson, M.; Blozis, S.; Bleibaum, R.; Cyrot, D. and Waterhouse, A.L. (2022). Wine Closure performance of three common closure types: chemical and sensory impact on a Sauvignon Blanc wine. *Molecules*,1**27(18)**: 5881. doi: 10.3390/molecules27185881. PMID: 36144619; PMCID: PMC9505717.

Chen, H.;Yang, J.;Deng, X.;Lei, Y.;Xie, S.;Guo, S.…Xu, T. (2020). Foliar-sprayed manganese sulfate improves flavonoid content in grape berry skin of Cabernet Sauvignon (*Vitis vinifera* L.) growing on alkaline soil and wine chromatic characteristics. *Food Chemistry*, **314**:126182

Colombo, R.C.;Roberto, S.R.;Nixdorf, S.L.;Pérez-Navarro, J.;Gómez-Alonso, S.;Mena-Morales, A.…Hermosín-Gutiérrez, I. (2020). Analysis of the phenolic composition and yield of 'BRS Vitoria'seedless table grape under different bunch densities using HPLC–DAD–ESI-MS/MS. *Food Research International.* **130**:108955. doi: 10.1016/j.foodres.2019.108955.

Dagadkhair, R.A. (2022). Development of probiotic food product from Manjari Medika grapes. Ph D thesis submitted to School of Food Technology, MIT-ADT, Pune.

Diouf-Lewis, A; Commereuc, S. and Verney, V. (2018). Biowastes from wine as natural additive of polyolefins: Thermo- and photo-oxidation efficiency. *Journal of Applied Polymer Science. Sci.*, **135(34)**: 46607.

Grassi, F.; Labra, M.; Imazio, S.; Rubio, R.O.; Failla, O.; Scienza, A. and Sala, F. (2006). Phylogeographical structure and conservation genetics of wild grapevine. *Conservation Genetics*, **7(6)**: 837– 845. https://doi.org/10.1007/s10592-006-9118-9

Jankar, J.J.; Pawar, V.N. and Sharma, A. K. (2019). Study on raisin quality and storability affected by coating of guar gum and glycerol combinations. *Journal of Pharmacognosy and Phytochemistry*, **8(3)**: 774-777.

Karibasappa, G.S. and Adsule, P.G. (2008). Evaluation of wine grape genotypes by National Research Centre for Grapes at their farm at Pune, Maharashtra, India. *Acta Horticulture*,**785**: 497-504.

Keast, D.R.; O'Neil, C.E. and Jones, J.M. (2011). Dried fruit consumption is associated with improved diet quality and reduced obesity in US adults: Natl. Health and Nutrition Examination Survey,1999–2004. *Nutritional Research*,**31**:460–7.

Kersh, D.M.E.;Hammad, G.; Donia, M.S. (2023). A comprehensive review on grape juice beverage in context to its processing and composition with future perspectives to maximize its value. *Food Bioprocess Technology*, **16**: 1–23. https://doi.org/10.1007/s11947-022-02858-5

Khan, ZS;Chatterjee, NS;Shabeer, T.P.A.; Shaikh, S. and Banerjee, K. (2020). Profile of triacylglycerols, phenols, and vitamin E of Manjari Medika grape seed oil and cake: Introducing a novel Indian variety. *European Journal of Lipid Science and Technology*,**122(4)**:1900356. doi: 10.1002/ejlt.201900356

Khiari, R.; Zemni, H. and Mihoubi, D. (2019). Raisin processing: Physicochemical, nutritional and microbiological quality characteristics as affected by drying process. *FoodReviews International*, **35(3)**: 246– 298. https://doi.org/10.1080/87559129.2018.1517264

Kohlbacher, C. (2021). Grape leather- the vegan wine fabric from Vegea. https://danandmez.com/blog/grape-leather/

Lacombe, T. (2012). Contribution à l'étude de l'histoire évolutive de la vigne cultivée (Vitis vinifera L.) par l'analyse de la diversité génétique neutre et de gènes d'intérêt, Montpellier, France.

Marangon, M.;Vincenzi, S. and Curioni A. (2019). Wine fining with plant proteins. *Molecules*. **24**:2186. DOI: 10.3390/molecules24112186

McGill, C.R.;Keast, D.R.; Painter, J.E.; Romano, C.S. and Wightman, J.D. (2012). Improved diet quality and increased nutrient intakes associated with grapeproduct consumption by U.S. children and adults: National Health and Nutrition Examination Survey 2003 to 2008. *Journal of Food Science*, **78(S1)**: A1-A4. https://ift.onlinelibrary.wiley.com/doi/epdf/10.1111/1750-3841.12066

McGill, C.R.; Wightman, J.D.; Fulgoni, S,A. and Fulgoni, V.L. III. (2011). Consumption of purple/blue produce is associated with increased nutrient intake and reduced risk for metabolic syndrome: results from the Natl. Health and Nutrition Examination Survey 1999–2002. *Am J Lifestyle Med*,**5**:279–90.

McGovern, P.E. (2003). Ancient Wine: The search for the origins of viniculture. Princeton: Princeton University Press. 373 p

Monge, M. and Moreno-Arribas, M.V. (2016). Applications of nanotechnology in wine production and quality and safety control. In: Moreno-Arribas, M., Bartolomé Suáldea, B. (eds) Wine Safety, Consumer Preference, and Human Health. Springer, Cham. https://doi.org/10.1007/978-3-319-24514-0_3

Michail, N. (2021). Israeli start-up Nutrilees has developed a proprietary process to remove the sour taste from wine lees, bringing a new antioxidant-packed, upcycled ingredient to the market.https://insights.figlobal.com/new-product-development/transforming-wine-waste-polyphenol-packed-powder

OIV. (2017). Distribution of the world's grapevine varieties. https://www.oiv.int/public/medias/5888/en-distribution-of-the-worlds-grapevine-varieties.pdf

OIV. (2022). State of the world vine and wine sector 2021. https://www.oiv.int/sites/default/files/documents/ENG_State%20of%20the%20world%20vine%20and%20wine%20sector%20April%202022%20v6.pdf

Parker, A.K.;García de Cortázar-Atauri, I.; Trought, M.C.; Destrac, A.; Agnew, R.; Sturman, A. and van Leeuwen, C. (2020). Adaptation to climate change by determining grapevine cultivar differences using temperature-based phenology models. *OENO One*, **54(4)**:955–974. https://doi.org/10.20870/oeno-one.2020.54.4.3861

Pawar, D.A.; Giri, S.K. and Sharma, A.K. (2021a). Development and performance evaluation of grape debunching machine for production of raisins and stemless table grapes. *Journal of Food Process Engineering*, **44(12)**.DOI: 10.1111/jfpe.13898

Pawar, D.A.;Giri, SK.; Sharma, A.K. and Kotwaliwale, N. (2021b). Effect of abrasive pre-treatment on drying rate of grape berries and raisin quality. *Journal of Food Processing and Preservation*, **45(9)**: DOI: 10.1111/jfpp.15746

Perestrelo Rosa, L.Y.; Santos Sónia A.O.;Silvestre Armando J.D.;Neto Carlos, P.;Câmara José, S. and Rocha Sílvia, M. (2012). Phenolic profile of Sercial and Tinta Negra *Vitis vinifera* L. grape skins by HPLC–DAD–ESI-MSn: Novel phenolic compounds in *Vitis vinifera* L. grape. *Food Chemistry*,**135(1)**:94–104.

Pudhir, A; Sharma, A.K.; Banerjee, K.; Jogaiah, J. and Somkuwar, R.G. (2018). Improvement in functional, rheological and sensory properties of low sugar ice cream by adding fine wine lees.*Progressive Horticulture*,**50(1-2)**: 118-123.

Sharma, A.K.; Sawant, I.S.;Sawant, S.D.; Saha, S.; Kadam, P. and Somkuwar, R.G. (2017). Aqueous chlorine di-oxide for the management of powdery mildew *vis-à-vis* maintaining quality of grapes and raisins. *Journal of Eco-friendly Agriculture*,**12(2)**: 59-64.

Sharma, A.K. and Aglawe, M.K. (2022). Addition of processed fine wine lees of Cabernet Sauvignon to improve nutraceutical properties of yoghurt. *Proc. of National Academy of Science India B: Biological Science*, **92(1)**: 141–147. https://link.springer.com/article/10.1007/s40011-021-01326-y

Sharma, A.K.; Upadhyay, A.K. and Somkuwar, R.G. (2020). Grape growing: Opportunities for better returns. *Progressive Horticulture*, **52(2)**: 134-143. DOI: 10.5958/2249-5258.2020.00018.4

Sharma, A.K.; Kumar, R.; Azad, Z.R.A.A. and Adsule, P.G. (2015). Use of fine wine lees for value addition in ice cream. *Journal of Food Science and Technology*. **52(1)**:592–596. DOI: 10.1007/s13197-013-1007-x

Sharma, A.K.; Dagadkhair, R.A. and Somkuwar, R.G. (2018). Evaluation of grape pomace and quality of enriched cookies after standardizing baking conditions. *Journal of AgriSearch*, **5(1)**: 50-55. https://doi.org/10.21921/jas.v5i01.11134

Sharma, A.K.andShabeer, A.T.P. (2019). Zero waste processing technology for high value products from Manjari Medika: a grape variety of ICAR-NRCG Extension Folder No. 37. https://nrcgrapes.icar.gov.in/Technical%20or%20Extension%20folders/'Zero%20Waste'%20processing%20technology%20for%20high%20value%20products%20from%20Manjari%20Medika-a%20grape%20variety%20of%20ICAR-NRCG.pdf

Sharma, A.K.; Somkuwar, R.G.; Upadhyay, A.K., Sawant, S.D. and Naik, S. (—) Production of quality and safe dried grapes. Pp 12. Extension Folder No. 9. https://nrcgrapes.icar.gov.in/zipfiles/NRCG-Quality%20and%20Safe%20Dried%20Grapes.pdf

Silva, F.V.M. and Van Wyk, S. (2021). Emerging non-thermal technologies as alternative to SO_2 for the production of wine. *Food*, **10**:2175. DOI: 10.3390/foods10092175

Yang, B.; He, S.; Liu,Y.;Liu, B.; Ju, Y.; Kang, D.;Sunn, X. andFang, Y. (2020). Transcriptomics integrated with metabolomics reveals the effect of regulated deficit irrigation on anthocyanin biosynthesis in Cabernet Sauvignon grape berries. *Food Chemistry*, **314**:126170. doi: 10.1016/j.foodchem.2020.126170.

Zhang, W.; Pan, Z.; Xiao, H.; Zheng, Z.; Chen, C. and Gao, Z. (2018). Pulsed vacuum drying (PVD) technology improves drying efficiency and quality of Poria cubes. *Drying Technology*, **36(8)**: 908– 921.https://doi.org/10.1080/07373937.2017.1362647

Chapter 18

Post-Harvest and Value Addition Technologies for Horticultural Crops

Dev Raj* and AK Pandey
Department of Post Harvest Technology,
Navsari Agricultural University, Navsari (GUJARAT) - 396 450, INDIA
**Professor and Head*

India has attained global supremacy in the production of fruits and vegetables, occupying the first and second positions, respectively. Horticultural crops including fruits, vegetables, flowers, plantation crops, spices and medicinal and aromatic plants have emerged as a major economic activity in India. Horticulture crops cover over 24.20 million hectares of area, which is approximately 11.25% of the total gross cropped area of the country and contributes 18-20% of the gross value of India's agricultural output (277.35 million tonnes production). India is the world's second largest fruit and vegetable producing country accounting for 88.48 million tonnes of fruit and 162.90 million tonnes of vegetables (NHB 2014). The country produces 41% of world mangoes, 23% banana, 24% cashew nut, 36% green peas, 10% onion and 30% cauliflower and accounts for 12% of world fruit and 13% of vegetable production (MOFPI, 2009). Owing to highly perishable nature, fruits and vegetables have very short shelf-life. During different handling and marketing operations there is 25-33% post harvest loss of horticultural produce. The major cause of post harvest loss is availability of poor infrastructure for post harvest management (PHM) and processing of commodities. These losses can only be minimized by proper handling, marketing and processing of the agricultural commodities. According to national level study conducted under AICRP on PHT, the post-harvest losses during different farm handling operations like harvesting, sorting, grading and packing accounts for about 13% ,during farm storage about 6% and during storage at godown, wholesale and retail level about 12% of the produce goes waste. Thus on an average, about one third of horticulture produce never reaches the ultimate consumer. Insufficient knowledge of pre and post-harvest operations and lack of proper facilities for handling like pre-cooling, grading, packaging, transport, storage, processing and marketing altogether compound the post-harvest losses for more than Rs 65,000 crores. The food processing industry ranks fifth in size in the country and employs 16 lakh workers which is 19% of the country's industrial labour. It accounts for 14% of the total industrial output with 18% of industrial GDP and 6.3% of countries GDP. Despite such good rank, there is only <2.4% processing in India as compared advanced countries (60-70%). So, post harvest technology can plays a vital role in India's prosperity. The status of horticultural crop production in India and Gujarat is presented in Table 1 to Table 4.

Table 1: Status of horticultural crops production in India

Status		Position
Area		24.20 million hectares
	Fruits	7.9 million hectares
	Vegetables	9.5 million hectares
	Others	3.3 million hectares
Production		
	Fruits	88.48 million tones
	Vegetables	162.90 million tones
	Flowers, Spices and Aromatic	5.66 million tones

Table 2: Status of Horticultural crop production in Gujarat state

Crop	Area (In 000' Ha)	Production (In 000' MT)	All India Rank
Fruits	349.9	7245	4th
Vegetables	515.9	9379.5	5th
Flowers (cut)	12.5	5063	6th
Plantations	23.1	129.3	NA
Spices	479.4	792.6	2nd

Table 3: Status of fruit crop production in Gujarat state

Fruit crop	Area (In 000' Ha)	Production (In 000' MT)	All India Rank
Papaya	17.8	974.0	2nd
Banana	64.7	3978	3rd
Mango	130.1	911.3	5th
Sapota	28.8	288.0	3rd
Citrus	39.2	409.1	5th
Guava	10.2	150.7	7th
Pomegranate	5.8	60.3	3rd

Table 4: Status of vegetable crop production in Gujarat

Vegetable crop	Area (In 000' Ha)	Production (In 000' MT)	All India Rank
Potato	61.1	1657	5th
Onion	62	1514.1	3rd
Brinjal	72	1236.3	4th
Tomato	38.8	978.4	7th
Okra	54.5	592.5	5th
Cauliflower	21.1	387.4	5th
Cabbage	28.2	553.6	5th

Fruits and vegetables are highly perishable and have very short shelf-life. During different handling and marketing operations there is huge post harvest loss of agriculture produce. Both qualitative and quantitative losses occur in horticultural commodities between harvest and consumption. Qualitative losses like loss in edibility, nutritional quality, calorific value and consumer acceptability of fresh produce are much more difficult to assess than are quantitative losses. Quantitative post-harvest losses in India estimated by different committees ranged between 25-33% depending upon the crop. The major cause of post harvest loss is availability of poor infrastructure for post harvest management (PHM) and processing of commodities. These losses can only be minimized by proper handling, marketing and processing of the agricultural commodities. According to national level study conducted under AICRP on PHT, ICAR (unpublished) , the post-harvest losses during different farm handling operations like harvesting, sorting, grading and packing accounts for about 13% ,during farm storage about 6% and during storage at godown, wholesale and retail level about 12% of the produce goes waste. Thus on an average, about one third of horticulture produce never reaches the ultimate consumer. This results in a considerable gap between gross food production and net availability. Insufficient knowledge of pre and post-harvest operations and lack of proper facilities for handling like pre-cooling, grading, packaging, transport, storage, processing and marketing altogether compound the post-harvest losses and wastage which in value terms accounts for more than Rs 50,000 crores. The status of post harvest losses in India is presented in Table 5 to Table 7.

Table 5: Status of Post harvest losses (India)

Name of fruits/vegetables	Post harvest losses (% of production)
Banana	12-14
Mango	17-37
Citrus (orange)	8-31
Guava	3-15
Apple	10-25
Pineapple	5-20
Grapes	23-30
Beans and peas	7-12
Brinjal	10-13
Cabbage	7-15
Cauliflower	10-15
Garlic	1-3
Onion	15-30
Potato	15-20
Tomato	10-20

Table 6: Status of Post harvest losses (%) at various stages

Crops	Harvesting (%)	Grading (%)	Transportation (%)	Marketing/ Storage Wholesaler (%)	Retailer (%)	Total loss (%)
Orange	3.25	0.75	1.25	1.20	7.50	13.95
Pineapple	1.73	0.54	1.95	2.66	2.37	9.25
Banana	Nil	Nil	10.00	5.00	7.00	22.00
Ginger	1.5	2.25	1.50	2.75	2.50	10.50
Tomato	0.75	3.75	11.00	2.50	7.25	10.50
Cauliflower	Nil	2.75	7.50	1.75	3.75	15.75
Spine Gourd	Nil	8.30	4.30	1.80	2.70	17.10
Pointed Gourd	Nil	5.40	7.50	1.90	2.10	16.90

Table 7: Cumulative post harvest losses of horticultural crops

Operation	Position
Post Harvest (Cumulative)	25-33%
Handling	13%
Farm storage	6%
Retailer/Godown/Wholesaler	12%

Post Harvest Management

Fruits and vegetables are characterized as the protective foods as they provide essential vitamins, minerals, dietary fibres, antioxidants and bioflavanoids. Accordingly, increased consumption of a variety of fruits and vegetables on daily basis is recommended for health benefits like reduced risk of certain form of cancers, coronary heart diseases, stroke and other chronic ailments. Although country is a major producer of these crops, many Indians are still unable to get their daily requirements of fruits and vegetables as per the recommended daily allowance of 280-350g fruits and vegetables per capita per day (Swaminathan, 1993). The main objectives of postharvest management and processing fruits and vegetables are i) to maintain quality (appearance, texture, flavor and nutritive value) ii) to protect food safety, and iii) to reduce losses between harvest and consumption. There are many interacting steps involved in any postharvest system. Produce after harvest is often handled by many different people, pretreated, pre-cooled, transported and stored repeatedly between harvest and consumption. Different post harvest operations takes place for management of horticultural crops. These are:

Cleaning & sorting, Washing & air drying, Grading, Pre cooling, Packaging, Ripening, Storage, Transportation and Market distribution

Cold storage of the horticultural crops can play an important role in minimization of the post harvest losses. The existing infrastructure available for storage of the horticultural commodity can store about more than 9-10% of the annual production. There are about 5101 cold storages in the country which can handle only about 21.7 million tonnes of the produce (DMI, 2006). The export of fresh fruit and vegetables from India in 2006-07 was estimated at 9.84 Lakh tonnes valued at Rs 2411.70 crores (Table 8). The numbers of cold storage units in different states of India is presented in Table 9.

Table 8: Status of storage and export of horticultural crops

Status	Position
Cold Store units	7091
Storage Capacity	21.7 million tones
Export of Fresh Fruits/vegetables	9.84 Lakh tones
Value	Rs. 2411.7 Crores

Table 9: Status of cold storage units in different states of India

Sr. No	State Name	Cold Storage No.	SL. No	State Name	Cold Storage No.
1	Andhra Pradesh	413	16	Maharashtra	555
2	Arunachal Pradesh	2	17	Manipur	1
3	Assam	35	18	Meghalaya	4
4	Bihar	304	19	Mizoram	3
5	Chandigarh	7	20	Nagaland	2
6	Chhattisgarh	97	21	Orissa	120
7	Delhi	97	22	Punjab	617
8	Gujarat	625	23	Rajasthan	157
9	Haryana	307	24	Sikkim	3
10	Himachal Pradesh	34	25	Tamil Nadu	165
11	Jammu & Kashmir	29	26	Tripura	14
12	Jharkhand	56	27	Uttar Pradesh	2209
13	Karnataka	192	28	Uttrakhand	30
14	Kerala	19	29	West Bengal	506
15	Madhya Pradesh	275			

Processing and Value Addition

The food processing industry ranks fifth in size in the country and employs 16 lakh workers which is 19% of the country's industrial labour. It accounts for 14% of the total industrial output with 18% of industrial GDP and 6.3% of countries GDP. So, food processing plays a vital role in India's prosperity. The Food Processing Industry sector in India is one of the largest sector in terms of production, consumption, export and growth prospects. Food processing industry broadly comprises primary processing category and the other categories embracing the secondary, tertiary and further stages of downstream processing.

In India majority of the produce is sent to the market for fresh trade and whatever is left as marketable surplus in the form of small and pittoo size, non-uniform and cull is utilized for processing for conversion into value added products.

The processing capacity in our country from more than 36871 licensed Food Processing Units is less than 2.4% of total fruit and vegetable production against 60-70% in many advanced countries. Further, the actual production of processed products stood at only 1.33 million tones with a less than 35% capacity utilization of the installed processing units. There are 70% units in home/cottage/ small scale sector and 30% units in large scale sector with capacity of 250 tonnes per annum and 30 tonnes/hr, respectively. Large scale contributes 70% of the production. Despite such a huge production of horticultural commodities there is only <2.4% processing and India's share in the world trade is around 1%. However, processing in advanced countries is too high as compared to India. Status of processing in India is presented in Table 10 and 11. India is the land of spices producing all varieties worth Rs. 3500 crores amounting to 25-30% of world production, which is processed for value addition and export. Tea, coffee, cashew, cocoa, sugarcane etc. are the major plantation crops in the country. India's principal crops accounted for 5-6% of the India's total export earnings. India is the world's largest producer of the tea. India is the largest producer (31% of world production), processor and exporter of the cashew kernels (48% of the world export). Consumer food comprises of group of products like chocolates, confectionery, cocoa products, soya based products, RTE foods, mineral waters, soft drinks etc. Among these, soft drinks enjoy the biggest share. Confectionery output is growing at the rate of 6-7 % and chocolate production at the rate of 10-15%. Mango, grapes, apple, onion, potato, green vegetables like okra, bitter gourd and green chillies were the major items of export. Similarly, India exported 9.53 Lakh tones of processed foods comprising of mango pulp, juices, concentrates, dried and processed vegetables, pickle and chutney, alcoholic and non-alcoholic beverages worth Rs 10583.41crores (APEDA, 2007) (Table 12 and 13).

Table 10: Status of Post harvest Management and processing

Status	Position
Fruits and vegetables Processing	<2.4%
Export of Processed Products	9.53 Lakh tones
Value	Rs. 10583.41 Crores
Processing capacity	3.85 million tones
Actual Production	1.33 million tones
Capacity utilization	Less than 35%

Table 11: Status of processing units in different states of India

Sr. No	State	Total no. of Processing units	SL. No	State	Total no. of Processing units
1	Andhra Pradesh	9359	16	Maharashtra	3113
2	Arunachal Pradesh	2	17	Manipur	18
3	Assam	1212	18	Meghalaya	18
4	Bihar	715	19	Mizoram	3
5	Chandigarh	23	20	Nagaland	12

Sr. No	State	Total no. of Processing units	SL. No	State	Total no. of Processing units
6	Chhattisgarh	1028	21	Orissa	875
7	Delhi	145	22	Punjab	2784
8	Gujarat	1924	23	Rajasthan	777
9	Haryana	650	24	Sikkim	18
10	Himachal Pradesh	171	25	Tamil Nadu	5186
11	Jammu & Kashmir	150	26	Tripura	55
12	Jharkhand	169	27	Uttar Pradesh	2216
13	Karnataka	1979	28	Uttrakhand	381
14	Kerala	1437	29	West Bengal	1600
15	Madhya Pradesh	754			

Table 12: Production of Processed Products in India

S.No.	Products	Production (,000 tones)
1.	Mango pulp	737.10
2.	Fruit juice	293.83
3.	Fruit nectars	125.70
4.	Dried fruits	84.8
5.	Preserves and candies	73.7
6.	Fruit Powder (flour)	15.4
7.	Mixed fruit pulps	3.3
8.	Juice concentrate	0.5
	Total	1334.3

Table 13: India's export of processed products

S.No.	Item	2016-17	
		Quantity (,000 MT)	Value (Crores)
1.	Dried and preserved vegetables	87.280	1088.55
2.	Mango Pulp	135.621	864.97
3.	Pickle & Chutney	180.820	942.72
4.	Alcoholic & Non-alcoholic beverages	231.332	2000.62
5.	Other processed fruits & Vegetables	351.83	3116.07
6.	Miscellaneous prepared products	283.265	2570.48
		1270.148	10583.41

Technologies Developed by Department of PHT at NAU

The Department of Post Harvest Technology (PHT) was established during 2004 under ASPEE College of Horticulture and Forestry, NAU, Navsari. The department teaches undergraduate courses and offers post graduate programs supported by well-trained faculty. At present two years M.Sc. and three years Ph.D. Degree Programmes are running in the Department of Post Harvest Technology. These degree programmes are formulated for developing competent Human Resource power for which significant job opportunities exist in this country. With advancement in production technology, the high yield / area of crops lead to large amount of marketable surplus of food grains and crop residues, demanding appropriate handling, processing, preservation, storage, marketing and utilization. The development of processing industries to preserve the perishable agricultural produce will not only improve economic and nutritional status of our population but it may help in employment generation in rural as well as urban areas of the country. This can be achieved by linking production, and post harvest technology in synergistic way. For this purpose, the department is equipped with excellent Fruit and Vegetable Processing Units for pilot scale testing of technologies, providing in-plant training and imparting community canning service to the students, farmers and enterpreneurers. The research achievements of the Department are enlisted here under;

i. Developed process for freeze drying of sapota slices for high quality dehydrated sapota chips and powder. High quality dehydrated sapota chips can be prepared by pre-freezing 5 mm thick sapota slices in freezers for 10 hours followed by freeze drying under vacuum of 760mm Hg at a temperature of 70°C for 12 hours.

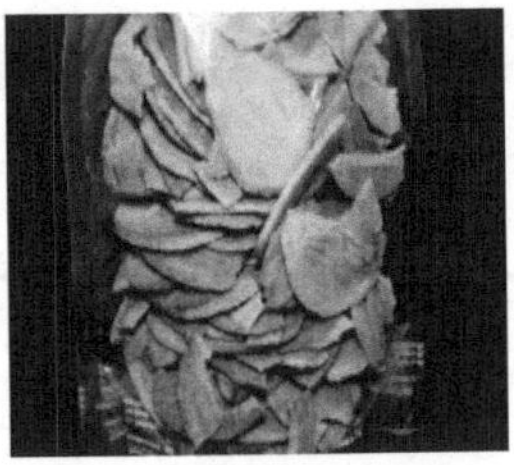

Chips

ii. Developed technology for utilization of banana peel for preparation of *sev*. Ripe banana peel can be utilized for preparation of *sev*. Ripe banana peel must be pre-treated immediately to prevent enzymatic browning by dipping in 2% salt (NaCl) solution along with 100 ppm ascorbic acid for 30 min. After pre-treatment, banana peel must be blanched, grind to make paste and mixed (30% ripe banana peel paste) with gram flour (70%) for preparation of fibre rich *sev*. The recipe for the preparation of ripe banana peel based *sev* comprised of 30 g ripe banana peel paste, 70 g gram flour, 2.5 g common salt, 1.5 g chilli powder, 0.75 g white pepper powder, 1.0 g turmeric powder, 2.5 g coriander powder and 5 ml edible oil.

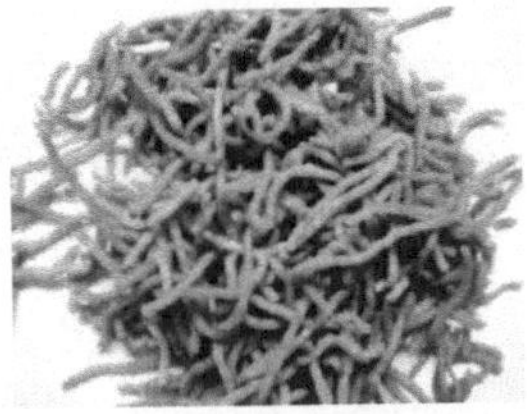

Sev

iii. Standardized method for extraction of 'Noni' (*Morinda citrifolia*) fruit juice. Noni juice can be extracted by treating crushed fruits with 0.10 % pectinase for 3 hours to get higher juice recovery. The juice after extraction must be filtered, pasteurized (96°C), packed in glass bottles followed by processing (96±1°C) for 30 min. The packed juice has storage stability for 12 months at ambient temperature.

Noni Juice

iv. Standardized the formulations for preparation of noni mango nectar from Noni juice. Noni juice can be utilized for preparation of blended noni mango nectar to increase the acceptability of noni juice. For preparation of blended noni mango nectar, blend 5% noni juice with 15% mango pulp by maintaining 16^0 Brix TSS and 0.3% acidity. The nectar after blending, filtered, pasteurized (96°C), packed in glass bottles followed by processing (96±1°C) for 30 min. The packed nectar has storage stability for 6 months at ambient temperature.

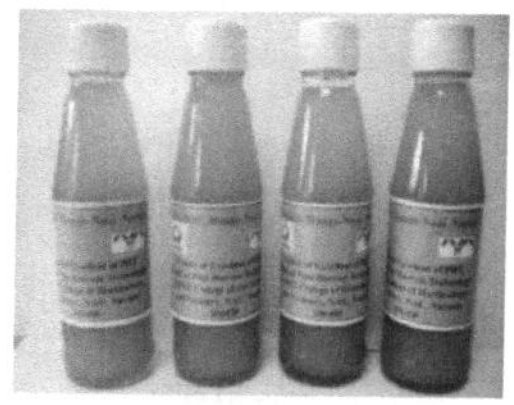

Noni Mango nectar

v. Developed technology for dehydration of onions rings. Red onions rings can be dehydrated by pre-treating onion rings with combination of 2000 ppm potassium meta-bisulphite (KMS) and 500 ppm citric acid for 15 minutes followed by dehydration at four stage dehydration temperatures (75, 70, 65 and 60°C for 2 hours, 2 hours, 1 hour and about 8 hours till drying, respectively).

Dried Onion flakes

vi. Developed technology for dehydration of okra slices. Okra slices can be dehydrated by pre-treating okra slices with combination of 1500 ppm KMS and citric acid @ 500 ppm for 15 minutes following two stage dehydration temperatures (75 for 2 hours and 65°C for about 10 hours till drying).

Dried Okra

vii. Developed technology for dehydration of cauliflower segments. Cauliflower cut segments can be dehydrated by pre-treating cauliflower cut segments with combination of 1500 ppm KMS and 1000 ppm citric acid for 15 minutes following four stage dehydration temperatures (75, 70, 65 and 60°C for 2 hours, 2 hours, 1 hour and about 7 hours till drying, respectively).

Dried Cualiflower

viii. Developed technology for preparation of Ready to Serve (RTS) beverage from banana pseudostem sap. RTS beverage can be prepared from blend of banana psuedostem sap and aonla fruit juice having 3.5% and 8% TSS respectively with the ratio of 90:10 could be stored up to six months in glass bottle.

Banana Pseudostem RTS

ix. Standardized technology for Processing of Banana Central Core Jam. The processors and house wives are recommended to prepare banana pseudo stem central core jam by replacing up to 50% fruits (mango, guava, papaya, pineapple) with central core. Mix fruit jam with central core is most acceptable combination as it not only reduces the production cost but also increase the fibre content of the jam without affecting jam quality.

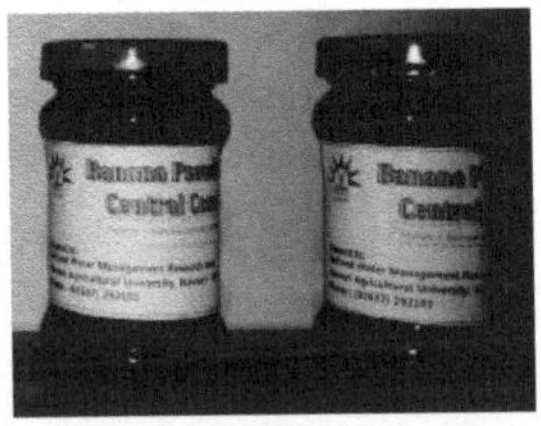

Banana Central core Jam

x. Standardized technology for Processing of Banana Central Core into candy. Candy from banana pseudo-stem central core can be prepared by giving blanching pre-treatment for 3 min followed by osmotic dip to cubes of central core in 70° Brix sugar syrup.

Banana Central core Jam

xi. Standardized technology for preparation of water melon rind candy by mixing 100 g sugar per 100 g water melon rind along with 0.20 per cent acid. The prepared candy was rated best on the basis of higher sensory score as well as nutritional composition.

Watermelon rind candy

xii. Standardized technology to utilize Mango peel and kernel powder for preparation of biscuits by using formulation of 5% mango peel powder, 7.5% kernel powder and 87.5% maida (F_7) and remained shelf stable during storage for three months in polypropylene bags.

Peel and kernel based biscuits

xiii. Standardized recipe for preparation of fruit bar from guava and papaya pulp (50:50).

xiv. Standardized recipe for preparation of mango bar containing desiccated coconut powder (2%).

xv. Standardized recipe for preparation of guava (cv. Lalit) nectar containing 16% pulp and 17 °Brix TSS and 0.3% acidity.

xvi. Developed technology for pre-cooling of sapota for reduction in losses and extension of shelf life. Sapota fruit can be precooled at 10 C in pre-cooling chamber with air movement of 263- 340-350 m^3/min for 5h can extends the shelf life of sapota fruits by 2 additional days

xvii. Standardized the recipe for the preparation of jam from the fruits of Palmyra palm. Jam from tender fruits of palmyra palm can be prepared by using pulp:sugar ratio (1:1.2) and addition of pectin 16g/kg of pulp and it also can be stored for six months at ambient temperature in glass bottle.

xviii. Standardized the recipe for the preparation of jelly from the *neera* of Palmyrah palm. Jelly from the *Neera* can be prepared by using pectin 13 g/kg of *Neera* and can be safely stored for six months. Recipe should be *Neera*: sugar (1:1.1), 0.5% acidity (50 g citric acid per kg of jelly) and pectin. Boil the mixture till 68°Brix followed by hot filling in to glass bottle.

xix. Standardized the recipe for the preparation of candy from the fruits of Palmyra palm. Candy from the fruits of Palmyra palm can be prepared by steeping the slices (5cm x 5mm) in sugar syrup having 65% TSS for 8 hours followed by drying of slices for 7 hours at 65°C and packed in PE pouches can be stored successfully up to six month at ambient storage.

xx. Standardized hot water dip treatment for eradication of fruit fly and extension of shelf life of mango fruits cvs. Kesar and Alphonso. Exporters are recommended to give hot water treatment at 50°C for 20 min to eradicate fruit fly infestation in Kesar and Alphonso mango for getting export quality fruits.

xxi. Developed process for utilization of dropped immature mangoes to prepare ready to eat pickle. Ready-to-eat pickle (*Moriya*) can be prepare from the immature dropped mango (*Marva*) through the process of washing, cutting and subsequently dipping in hot water (50 °C) for 5 min, 2% brine solution for 5 min and 5% acetic acid (vinegar) for 5 min, and finally mixing with the groundnut oil (2.5%)-spices mix and packing of in 75 micron HDPE bag.

xxii. Developed protocol for shelf life extension of the mangoes. Kesar mango fruits can be pre-cooled at 10 C in pre-cooling chamber with air movement of 300-350 m^3/min for 8h and pack them in 75micron polypropylene bag and store at 11±1°C with 90-95% RH. This would delay the initiation of ripening process up to 25days of storage and extend the shelf life up to 35 days.

xxiii. Standardized the technique for colour extraction from *Butea monosperma* flowers for preparing herbal *gulal. Butea monosperma* flower could be used for colour dye extract using 50% methanol water based v/v solution at 60°C temperature and 4 hr. process time. The extracted colour matter dye could be use for production of herbal '*gulal*'.

xxiv. Optimized the level of temperature and KMS for processing of ripe banana puree at pilot scale. Addition of 250 ppm ascorbic acid at the time of milling with 750 ppm potassium matabisulphite is recommended to get good quality banana puree. Puree pasteurized at 90 °C temperature for 10 min, filled hot in to the sterilized tin can, followed by seaming and processing at 100 °C temperature for 30 min and cooling. Puree has storage stability up to 6 months.

xxv. Optimized the level of TSS and anti-caking agent in spray solution for preparation of powder from ripe banana at pilot scale. For preparing spray dried banana powder, use 10 °Brix spray solution of banana puree after adding 15 % Maltodextrin as anti-caking agent. Spray should be done by keeping feed flow rate 35.0 kg/hr, feed temperature 70 °C, inlet temperature 170 °C and outlet temperature 100 °C for minimizing the sticking issue of banana puree in the inner chamber of spray drier.

xxvi. Technology for preparation of organic liquid fertilizer from banana pseudostem has been developed at NAU and is internationally patented. This organic liquid fertilizer is rich source of all macro r and micro nutrients as well as plant growth hormones; which enhances the productivity of crop and reduce the use of chemical fertilizers. It can be use through micro-irrigation method as well as as can be use as foliar spray

Future Strategy

1. Crop productivity and quality maintenance through selection, breeding, biotechnology and integrated pre - and post harvest management.
2. Minimization / prevention of wastage and quality loss through development of cool chain technology.
3. Minimization of wastage and quality loss by reduction in the numbers of intermediary produce handling stages and agencies.
4. Stabilization of produce supply and prices by maintenance of proper procurement and distribution chain.
5. Increase in the farmer's share of the consumer price
6. Expansion of raw material availability and up gradation of its quality.
7. Harmonization of various food laws.
8. Rationalization of taxes and levies.
9. Investment and improvement in infrastructure like rural roads, rural electrification, waste land development, cold chain etc.

Expected Output

Health, fitness, wellbeing, nutrition, convenience and freshness are likely to drive the Indian Food Processing Industry in the years ahead, as in the rest of the world. So for the fulfillment of these needs; health oriented nutraceuticals and functional foods, minimally processed foods, fresh cut, pre-prepared vegetables and salads, heart healthy, diabetic friendly foods and food supplements, low sugar / salt / fat / cholesterol foods, organically grown foods and other kindred food products may be the drive foods of the future. Convenience foods like instant mixes, ready – to – cook, ready-to-bake, ready -to – eat and ready-to-drink foods and beverages already becoming popular may soon become the order of the day.

Chapter 19

Protected and Precision Farming Technologies for Enhancing the Productivity of Horticultural Crops

M.C. Singh*, P. K. Singh, Jugendra Kapasia, Abhishek Mishra and Arun Kumar Dwivedi

Centre for Protected Cultivation Technology, ICAR-Indian Agricultural Research Institute, New Delhi-110012.

**Email :mamsingh@gmail.com; mam.singh@icar.gov.in*

Precision and protected agriculture envisages the provisions to protect the crops against extreme temperatures, wind, rain, hailstorms, birds and inspect-pests for sustained plant growth, yield, and quality of horticultural crops. A greenhouse (polyhouse) comprised of its design, shape, the orientation, the structure, its foundation, cladding material, and the ventilation it can permit together with the proper function of technical equipment needed to control the climate inside the structure. The system must protect plants against low and high temperatures, wind, rains, hailstones, birds, and insects allowing modern work routines and logistics in the production chain. The incident solar radiation (the visible and adjacent portions of the infrared and ultraviolet ranges of the spectrum) passes through the plastic/glass roof and walls and is absorbed by the floor, earth, and contents, which become warmer and re-emit the energy as longer wavelength infra-red radiation. The cladding material used for greenhouse walls do not transmit infrared radiation, so the infrared cannot escape via radiative transfer. Irrigation methods influences the crop growth and accumulation of assimilates into final economic yield that reflects the water and nutrient use efficiency. Water productivity is their in terms of fresh fruit yields per unit area per unit applied fertigation. It has been already worked out that a water saving of 27% over drip, 33 % over sprinkler and 42 % over flood irrigation can be saved in rain pipe irrigation systems, producing more yields (15 %, 21% and 145%, respectively).Therefore, among the factors that influence the crop growth are, temperature (20+2 ^{0}C), relative humidity (60-65%) and light irradiance in terms of photosynthetically active radiation (PAR) @ 160-220 μM can best acclimate and sustain the growth to promote maximum photosynthesis.

Climate Management

Greenhouse cultivation is widely spread all over the world to produce quality and high value vegetables, flowers and fruits, seedlings/plants, herbs and or ornamentals. Besides, the structural design, prevailing climate, most important factor affecting plant growth is solar radiation governing plant growth from photosynthesis. Characteristics of the cladding material that can affect the level and quality of the transmitted radiation are of primary concern for greenhouse cultivation. Cladding materials need to be anti-drift, anti-dust, uv-resistant and >98% transparent, relates to their condensation characteristics. Such problems can be addressed by using a cladding material equipped with surface-active additives to create an anti-drop property and cooling additives to provide a shading effect. Not only it provides shades but also reduce the temperature values (- 4 ^{0}C), UV radiation (-26 W/m^2) and increase relative humidity (18-22 %), significantly. The highest advantage being the moderate values (180-260 μmol sec^{-1}m^{-2}) of photosynthetically active radiation (PAR) resulting in an increase in the productivity by 17 % in winter and 21% in summer. Light irradiance is an important factor in crop regulation especially in photosensitive crops. Vernalization from temperature and light is going to play a crucial role in having a possibility of growing 5 crops without compromising the yield and quality in an year's time. However, this control and regulation is against prevailing higher peaks of temperature, light and relative humidity which keeps a pace with enhanced water, nutrient and space use efficiency in polyhouse farming while growing indeterminate and high value crops.

Soil and Plant Protection

Plastic mulching of different colored films composed of degradable and bio-degradable polymer can be a boow for transient crop protection against input stress and mitigation of effects of hot and cold weather. Wherein water loss can be saved up to 27% and temperature sustained by 7^{0}C in summer and 4^{0}C in winter. An extra thin sheet of plastic film up to 15μ does not resemble traditional types of thick (25μ) plastic mulches, organic mulches, such as straw or leaves. Due to an advent of mechanical ease convenient rolls can be applied to the length of field rows to seal the upper layers of the soil. Crops are allowed to grow up through holes that are cut into the plastic covered beds. A wide array of plastic mulch products varying in color, thickness, dimensions, plastic type, and intended purpose as per crop requirements. Mulch materials and techniques significantly affect the microclimate around the crop canopy by changing the radiation budget of the soil top, soil water transformation, evapotranspiration (ET), and water use efficiency (WUE). WUE (kg/Ha-mm) was calculated as economic yield (kg/h) divided by the total ET (mm) from sowing to harvest. Unlike other field crops its evident that the plastic mulches are used in the cultivation of horticultural crops to modify the soil temperature and moisture regimes, control weeds, deter the immigration of insects, and possibly alter the photobiology of the plant. Not only this, but function of plastic mulch can also insulate and maintain a consistent temperature and humidity of the soil, preventing evaporation of moisture from the soil, minimization of seed time and harvest, prevent weed growth, and to prevent erosion. These microclimate factors strongly affect the soil temperature and soil moisture in the root zone, which, in turn, may influence plant growth and productivity. In technological verification of agricultural films and their optical properties, the experiments have been takes in different crops i.e., in polyhouses high value flowers (chrysanthemum, lilium and rose etc) and vegetable crops (cucumber, tomato and capsicum) and under mulching in open fields selected vegetables (broccoli, cabbage, cauliflower, lettuce, marigolds, tomato, potato and brinjal etc.) in conjunction with drip irrigation. The production technology of big-fruited tomato & cherry tomato, cucumber and capsicum has been developed and standardized for naturally ventilated greenhouse conditions and the yield levels are exceptionally high, attainable with superior quality.

Structure, Design and Profitability

Three crops of gynoecious parthenocarpic cucumber can be grown over duration of 10-11 months under naturally ventilated greenhouse conditions with productivity ranging between 120-130 t/ha. This technology eliminates stresses due to biotic and abiotic factors with minimum use of pesticides. The technology is highly remunerative for the growers near peri-urban areas of the country that can access and target niche markets as off-season produce which is not possible under outdoor cultivation. Year round production of high value vegetables namely, tomato, cucumber and capsicum with a high productivity (15-17.5t, 12-15t and 7-10 t fresh fruits per 1000m^{2}area) and profitability (at a cost benefit ration of 1:2.5, 1:2.5 and 1:2.0 respectively), under naturally ventilated polyhouse can be useful venture for small and marginal farmers. The protected cultivation of flower crops is being practiced as a small scale and the scientific Institutions and flower industry has prioritized the cut flower production with a higher economic scale, range of flowers and their products, suitable varieties, quality control and certification, cold chain management and transit facilities. However, the contribution in relation to flower trade is significant. Resultantly, flower crops namely, rose, chrysanthemum, carnation and gerbera are highly remunerative while cultivated in climate-controlled greenhouse and fetch handsome income at a cost benefit ratio of 1:3.6, 1:6.2, 1: 2.8 and 1:3.7, respectively.

Off-season Growing Strategies

Plastic low tunnels provide the best way for off season cultivation of cucurbitaceous vegetables during winter season by modifying the micro-climate around the plants. Low tunnels also offer several advantages like protection of the crop from frost, hails, and crop advancement from 30-40 days over their normal season of cultivation. This technology is gaining an importance where the low temperature during winter season is a severe constraint for cultivation of entire group of cucurbits and offers an advantage to maximum returns over the prevailing high market price for off-season produce. Seedlings of the required cucurbitaceous vegetables are raised through plug tray nursery raising technology. Thereafter, seedlings are transplanted on raised beds having drip irrigation facility. After transplanting the seedling individual rows of the crop are covered by making low tunnels at a height and width of 45-60 cm. Mostly, transparent plastic sheets of 20–30-micron thickness is used for making the low tunnels. Plastic low tunnels can be made manually and mechanically both. This technology is highly suitable for those areas where night temperature during winter season is going below 8°C for a period of 30-40 days (December-January months) and it is not possible to grow these vegetables in open fields. It is estimated that this technology offers the cost benefit ratio of 1: 3.0 -3.5 for summer squash, 1:2.5 – 3.0 for bottle gourd and bitter gourd and 1: 2.5-3.0 for musk melon.

Walk-in tunnel technology of growing vegetables is on an enormous increase under North Indian plains, there is severe winter during December and January months. An off-season cultivation of cucurbitaceous vegetables like summer squash, bottle gourd, cucumber, tomato and french bean etc can be taken up during severe winter using different protected structures. Farmers can earn higher profits due to high price for off-season produce in the market. Walk-in tunnels, which is a temporary structure made up of GI pipes and transparent polythene, can be used for off-season cultivation of vegetables like summer squash, bottle gourd, cucumber etc. during winter season (December-January) to fetch high price for off-season produce and to earn higher profit per unit area. This structure is very useful and profitable for vegetable growers. These temporary plastic structures are erected over the crop during the peak winter months of December to mid-February and thereafter, the structure can be removed from the crop. The plastic sheet can be reused for 2-5 years. Walk-in tunnel are suitable for growing off-season vegetable under North Indian plains and lower hills.

Hi- Tech Nursery Raising

Availability of quality seedlings of vegetables and flowers at right time is a challenge next after best quality seeds. Most of the farmers in India prepare seedlings in soil under open field conditions where biotic and abiotic stresses are major constraints in raising healthy and time-bound seedlings. Among the biotic stresses, soil borne fungus, nematodes and different viruses are the major limiting factors in growing healthy seedlings of vegetables like tomato, chili, sweet pepper, etc. Economic returns from these crops can be greatly enhanced if grown from healthy and disease-free seedlings any time in the year. The technology is aimed at raising healthy and season-independent seedlings. Not only the quality of seedling is achieved in shortest possible time but it avoids an establishment period of 2-3 weeks after transplanting resulting in an early produce for the market at reasonable price. Where the mortality of the seedlings is almost zero. This technology can give a pace to cater the increasing demand for quality and disease-free seedlings in and off-season. A greenhouse nursery of 500 m^2 area can produce 12-15 lakhs of seedlings in six batches per year with a cost of benefit ratio of 1:3.0. This can be adopted as self-employed enterprise by agri.-graduates or progressive farmers for enhancing their incomes.

Grafted Seedlings Against Biotic and Abiotic Stress

Grafting of vegetable and ornamental seedlings/saplings is a unique horticultural technology practiced for many years in East Asia to overcome issues associated with intensive cultivation using limited arable land. This technique was first practiced by grafting watermelon (*Citrullus lanatus*) onto pumpkin (*Cucurbita moschata*) rootstock in Japan and Korea in the late 1920s. A serious crop loss caused by soil-borne diseases aggravated by successive cropping was avoided by production of vegetables with grafted seedlings. In many fruit-bearing vegetables such as watermelon, cucumber, melon, tomato, eggplant and pepper, the use of grafted seedling has become increasingly popular. Grafting is an environment-friendly approach which is used to control soil borne diseases and increasing the yield of susceptible cultivars. This technique is eco-friendly for sustainable vegetable production and by using resistant rootstock, it reduces dependence on agrochemicals. To induce resistance against low and high temperatures, grafts are generally used. Grafting increases the yield and promotes biotic/abiotic stress tolerance. Grafting is also used to induce tolerance to abiotic stresses *viz.* flooding, drought and salinity. In Japan (92%), Korea (98%) and China (20%), major share in watermelon production is from grafted seedlings. In Europe, Spain is leading in grafted seedlings production with 129 million grafted seedlings followed by Italy (47 million grafted seedlings) and France (28 million grafted seedlings) (FAO 2009). Grafting as a technology for the commercial production of vegetables was later on adopted by many countries in Europe, Middle East, Northern Africa, Central America and other parts of Asia.

In India, vegetable grafting work was first started in ICAR-IIHR Bangalore by Dr RM Bhatt and his associates. Their work was on identification of rootstocks for waterlogged conditions. For this purpose they have imported semi-automated grafting machine. IIHR Bangalore organized first ever short course on vegetable grafting during the year 2013.Later on at many places work on vegetable grafting was started in a scattered manner. NBPGR regional station, Thrissur, Kerala have done work on Cucurbit grafting by taking *Momordica cochinchinensis*, a dioecious plant. The female plants were grafted on to the male plants to increase its production. Graft success was 98%. SAUs and ICAR institutes like CSK, HPKV, Palampur, ICAR-IIVR, Varanasi, ICAR-CAZRI, Jodhpur also initiated work on grafting and identified more than 22 rootstocks of brinjal, chilli, tomato and cucurbits for importing resistance to bacterial wilt and nematodes. Some private players are also involved in vegetable grafting. One of them is 'VNR Seed Private Limited' in Chhattisgarh which is supplying grafted brinjal seedlings resistant to bacterial wilt to farmers. The other seed company is Takii Seed India Private Limited and very recently CCS, HAU has also established a vegetable grafting nursery in public private partnership with VNR Seeds at Hisar, Haryana. Recently, Center for Protected Cultivation Technology, ICAR- Indian Agricultural Research Institute, Pusa,New Delhi has started working on evaluation of different rootstocks and their suitability for vegetable grafting and in near future grafted plants will be made available to the farmers.

Irrigation and Fertigation Strategies

Different irrigation modules namely, sprinkler, rain pipe and drip irrigation systems are in use in agriculture. Drip fertigation systems have shown an outstanding results with a 40% water saving and 65 % nutrients save in majority of horticultural crops. Rain pipe irrigation systems have further enhanced yield by 27% over standard method of irrigation (drip) and 43 % over the crops irrigated by flood irrigation. Furthermore, low pressure drip irrigation system requires comparatively less pressure than normal pressurized irrigation system and is suitable for small land holdings, greenhouses and nursery. This system does not require regular supply of electrical energy as the hydraulic water head of about 2 meter is sufficient for irrigating an area of about 500-1000 m^2. This system has high irrigation efficiency (greater than 90%) and is easy to install, operate and maintain. The system is consisted of the different components including a 500 liter water tank attached with 32 mm diameter LDPE main line pipe of about 30 meter, 120-mesh/130 micron disc/ screen filter along with the tank connectors and approximately 500 meter length of lateral drip line (pipe of 6/12 mm diameter fitted with in-line dripper of 0.65-1 lph discharge capacity) spread all over the beds. The cost involved for a unit area of 500 m^2 composed of an installation value at Rs10, 000/=. The system offers water saving of 30-50 % and is suitable for a large number of vegetable crops, flower and fruit crops. Annual production cycle of the crops may sustain a profit of the ` Rs 25,000/= per unit (500 m^2) and a cost benefit ratio of about 1:2.00.

Solar Powered Greenhouse Cultivation

A 200 square meter solar energy operated greenhouse equipped with fan-pad cooling system and low pressure drip installed with photovoltaic power generation capacity of 2.6Kwp. The system consists of solar module mounting, battery, solar PCU and wires could run two exhaust fans each of 500 watts and 1.0 HP pump for evaporative cooling and drip irrigation in a crops, capsicum and cucumber under soil-less system. The energy budgeting was calculated as total energy consumption in the greenhouse and varied from 300-800 Kw hrs on monthly basis. The maximum energy consumption of 800 Kw/hr. being in June. The total annual energy cost was estimated as Rs 24,400/ to meet the energy requirement of 6100 Kw hr for a 200 m2 evaporative cooled greenhouse. Growing of high value crops of vegetable (cucumber and capsicum) and flowers (carnation and chrysanthemum) have been tested with an input use efficiency (62% water and 27% fertilizers) as compared with the normal greenhouse crops.

Quality Seed Production

Most of the vegetable and annual flower crops are propagated by seeds and it is now well established fact that the seed is the most vital input for commercial crop production. Earlier, the cost of seed constituted only a fraction of the total cost of inputs involved in raising the crop, but nowadays seed is contributing a substantial amount to the total cost of cultivation. The basic reason behind this fact is that the use of quality seeds and hybrids has increased sharply during last few decades. The reason for the high cost of hybrid and quality seed is due to the poor seed yields, when produced under open field conditions. The production of genetically pure, healthy and disease-free seeds in vegetable and ornamental seed crops for commercial cultivation is a challenging task. Conventional practices of seed production in open fields' encounter common problems like lack of sufficient isolation, problem of virus free environment, incidence of insect-pests and diseases. Compared to conventional practices, protected cultivation can fetch higher seed yields with better quality. Moreover, incidence of insect-pests and viruses are the most devastating problems for quality seed production in most of the vegetable and ornamental crops in open fields, and if the insect vectors are checked by using protected structures, this problem can be solved with less application of pesticides.

An innovation applicable in a very practical manner is likely to get popularity to the assumed potential, resulting in significant profit to the stakeholders. Seed production under protected conditions is a best example of such innovation. This concept has shown significant results not only in quality seed production but also in increasing the per unit seed yield in different vegetables and ornamental crops. The protected cultivation models has proved to be highly economical and profitable with limited resources and manpower to provide best quality seeds of numerous vegetables and ornamental crops cultivated across the globe.The net house can be retained for 3-4 years with minor repair, thus the returns on seed production increases in the following years. The estimated cost benefit ratio is 1:1.6.The seed yield and quality is always better in seed production programmes undertaken in protected conditions. Semi-climate controlled greenhouses were found to be ideal for indeterminate type hybrids/varieties of standard tomato, cherry tomato, sweet pepper and parthenocarpic cucumber,bittergourd, melon, summer squash etc. The seed yield of such crops can be increased 2-3 times compared to the open field conditions. Besides, the quality of seed is also superior. Similarly, zero energy naturally ventilated polyhouses have also been found equally suitable for hybrid seed production of high value vegetable and ornamental crops. Here, the seed yield is also usually 2- 3 times more than open fields, but the cost of seed production could be reduced to almost half of the seed produced under semi-climate controlled greenhouse conditions.

Above discussed strategies of protected and precision farming can be explored for enhancing the productivity of horticultural crops in various ways.

References

Anonymous. (2017). Greenhouse cultivation of vegetables. CPCT, Indian Agricultural Research Institute, Pusa, New Delhi.

Anonymous. (2019). Protected Agricultural Technology for Hi-Tech agriculture under changing climatic conditions (in Hindi), DKMA, Indian Council of Agricultural Research, Pusa, New Delhi

NAAS (2010). "Protected Agriculture in North-West Himalayas" Policy paper no.47,National Academy of Agriculture Sciences New Delhi.pp16

Patel, Neelam, P. K. Singh, M.C. Singh, Ramesh Kumar, D.P. Singh and Anil Kumar "Design of recharge structure and sprinkler system for ground water recharge and vegetables production" in 9th International Micro Irrigation Conference (9IMIC) from 16 to 18 January, 2019 at Aurangabad p-100.

Patel, Neelam, Mairaj Hussain, P K Singh, M C Singh, Gaurav Singh, Sumit Pal (2019). Development of vertical farming technology for Peri-urban & Urban Horticulture (Abstract No. 1289) in XIV Agricultural Science Congress from February 20-23, 2019 at NASC, Pusa, New Delhi, India

Singh, B., Singh, A. K., & Tomar, B. S. (2010). In Peri Urban: Protected cultivation technology to bring prosperity. Indian Horticulture, 55(4), 31–32.

Singh, B., Tomar, B. S. & Thakur, S. (2010). Quality seed Production of Parental lines of Pumpkin (C. *moschata)* under Insect Proof net house. Acta Horticulturae (ISHS), 871, 275–278.

Singh, B., Singh, A., & Kumar, M. (2012). Pari nagriya Shetroo me: Aiise hogi sabjiyo kee sanrakshit kheti. Phal Phul. 33(3), 3–6.

Singh, H, Singh, M C, Singh, PK, KP Sahana and Mishra, I.M.(2021).Hi-Tech cultivation of capsicum. Indian Horticulture, pp.3-9 March-April -2021

Singh P.K. & Lal, S.K. (2018). Seed Production of vegetables and ornamental crops under protected conditions. In training manual of International training on Seed Production and quality evaluation.Jan14-28,2018. Division of Seed Science and Technology,IARI,New Delhi

Singh,P. K., Selvakumar, R, Singh, MC, Bhupal, HS, Jugendra Kapasia and Gautam R. (2021). Vegetable grafting for combating stresses and to increase productivity. Indian Horticulture, pp.43-48 March-April -2021.

Chapter 20

Floriculture: Blossoming Entrepreneurial and Career Opportunities in India

Alka Singh
Principal and Dean, ASPEE College of Horticulture
Navsari Agricultural University, Navsari-396 450, Gujarat
email: alkaflori@nau.in

India, with its diverse climatic regions and a rich floral biodiversity, is a heaven for the floriculture industry. Historically, Indians have cherished flowers for their aesthetic appeal and have integrated them into their religious and cultural traditions. Today, this passion for floriculture is gradually transforming into a promising entrepreneurial sector. Floriculture in India has evolved to the present form of lucrative flora Industry with the growing technology inputs, knowledge and skill up-gradation along with the global trade re-organized scenario and growing demand in national and international markets. Floriculture, the cultivation and management of flowers and ornamental plants, is a rapidly growing industry with vast career opportunities with high entrepreneurial prospects. The potential for floriculture in India, when harnessed appropriately, holds a bright prospect for becoming a significant contributor to the country's economy and a source of job creation.

At global level, flora business is very diverse and includes the production of cut flowers and cut foliage, flower bulbs, potted floweringas well as foliage plants and bedding plants. Besides, landscape gardening is a huge industry in itself is interrelated with it. Global floral production value is estimated at USD 54.59 bn. Europe, America and Asia account for more than 90 percent of the total world trade in floriculture products. The Netherlands has been continuously dominating the world floriculture industry from decades and account for major floriculture trade. However, Colombia, Kenya, Ecuador and Ethiopia have been gaining momentum in flower trade. In Asia, India, China and Thailand are moving progressively and emerging leading countries for floriculture produce. Asia-Pacific countries are the main suppliers to Japan and Hong Kong. African and other European countries are the principal suppliers to Europe's main markets, and the supplies to the United States is mainly catered by Colombia and Ecuador.

India is paving its way towards significant growth in the flower sector as Indian fresh flowers and plants are now been accepted while dry flowers contribute noticeably in the international market. Floriculture which was merely restricted to growing of traditional flowers in our country has emerged to a modern hi-tech agro-based industry. The popularity and commercialization of flowers is rapidly increasing as depicted from the magnificent floral decorations. Further, value added floral products like fresh flower craft and dry flowers craft are in much demand in both national and international market. Thus, besides cultivation of flowers, floriculture has an immense potentiality for youth to choose it as a career with a lot of entrepreneurial potential that is still not been fully explored.

Present Status of Flower Sector in India

Floriculture in India is now considered a lucrative commercial venture. It provides employment opportunities to many and contributes significantly to the economy. The Indian floriculture market size reached INR 231.7 Billion in 2022. Looking forward, It is expected the market to reach INR 460.6 Billion by 2028, exhibiting a growth rate (CAGR) of 13.1% during 2023-2028. In our country, flowers are grown in around 3,22,000 ha land, with the production of loose flowers around 21,52,000 MT and that of cut flowers of 8,28,000 tons (2020-21), according to the NHB database. This, however, excludes many flori-based industries like nursery business and dry flowers scenario due to unavailability of data although bedding plants and pot plant industry forms the major part of the world flower trade (>50%).

The main states involved in floriculture are Karnataka, Tamil Nadu, West Bengal, and Maharashtra.The Indian floriculture market is largely driven by the growing demand for flowers for religious purposes, decorative needs, and events like weddings and festivals. In addition, there's a rising trend of using flowers for personal adornment and gifting. Further, more than two-thirds of the area under floriculture is devoted to production of traditional flowers like marigold, jasmine, roses, chrysanthemum, tuberose etc. Several export-oriented enterprises have come up in the country, and more than 25,000 ha land is under protected cultivation.

Value addition in flowers through dry flowers, essential oils, tinting, natural dye, floral arrangements are gaining momentum and popularising in our country. Other associated activities of flower growing in India include the dry flower industry and the essential oil industry. India is the fifth largest exporter of dried flowers, and second largest exporter of dried foliage in the world accounting for around 7 percent of world exports in dry flowers and foliage. The main export markets for India's dry flower industry are USA, The Netherlands, UK and Germany.

There is a vast opportunity in value added floriculture trade at entrepreneurial level for rural and urban people including women besides growing and selling flowers. Value-adding is one such strategy that is also a logical extension of many farm businesses and can exploit underutilized farm resources for increased returns. Value-added products offer higher returns, new markets create brand recognition and add variety to farm operations. Small-scale farmers and young entrepreneurs can set up small enterprises where value addition in the form of fresh and dry flower crafting can be taken up. Value-added products can significantly enhance growers' businesses, and offer them a steady income even during the off-season.

Floricultural Exports

In view of exports, an exponential growth of floriculture products was being observed in a decade before 2006-7 when the exports reached to Rs 649.6 crores. But since then, a down fall to Rs 340.14 crores (2007-8), Rs 368.81 crore (2008-9) and Rs 294.46 crore (2009-10) was been witnessed owing to the downfall on the world economy which gradually improved with steady increase from the year 2011-12 and after almost a decade, the floricultural exports crossed the mark 500 crores since 2016-17, with around Rs 541.61 cr in 2019-20, Rs 575.98 cr in 2020-21, Rs 771.41 cr in 2021-22 and Rs 707.81 cr in 2022-23.These statistics reflect a positive trajectory for the floriculture industry. With the technical collaborations from foreign companies, the Indian floriculture industry is poised to increase its share in world trade.

Establishment of a large number of export oriented cut flower units has given recognition to Indian flowers in the international market and the growing dry flower industry that has steadily been contributing to 60-70% in the exports. The major importing countries were U S A, Netherland, U Arab Emts, U K and Germany. Thus, there exists the great potential and vast opportunities for export of floricultural products from India.

Commercial aspects of Floriculture for Career Option

There are many commercial aspects of floriculture and each aspect is capable of good income generation and has entrepreneurial potential. The breadth of entrepreneurial opportunities in floriculture is vast and encompasses areas such as flower farming, wholesale floristry, retail floristry, flower arranging and decor, flower exports, and technology-driven flower businesses. Each of these areas offers unique entrepreneurial potential. Some of the floriculture trade aspects for choosing of career option are given below:

1. Cut flower cultivation/consultancy/greenhouse management-India's various agro-climatic conditions are conducive for the cultivation of a wide variety of flowers, both exotic and indigenous. Entrepreneurs can venture into the production of cut flowers, loose flowers, or potted plants. As the demand for both indigenous and exotic flowers grows, there is a rising need for professional flower growers or floriculturists. These professionals cultivate a variety of flowers for different markets - loose flowers for religious or decorative use, cut flowers for arrangements, or potted plants for home and office decor.
2. Nursery Industry-It caters to supply of ornamental plants, bedding plants, plant rental service, seeds and bulbs etc. Bedding plants and pot plants account for more than 50 percent share of the total floriculture global trade. The nursery business in terms of ornamental plants, although un-estimated is vast in India. Plant nurseries fall into three main categories: the production (wholesale nursery), the growing-on nursery and retail nursery. Further, nursery based on rental plant services and the mail-order nursery also exists. Wholesale nurseries propagate plants, grow them in bulk and sell their plants to growing-on, retail and mail-order nurseries, as well as

to landscapers and landscape contractors or architects. The business of whole sale nurseries should focus on innovation i.e. new varieties and introduction of new plants, specialization (propagating of particular category and large quantity) and meeting of market demands through good market study, knowing market demand, taking the orders in advance and maintaining customer loyalty.

Retail nurseries should focus on defining their customer base, establishing good customer relations and booking the orders from the whole salers as well as linking with the retailers in advance. High-profit items for retail nurseries include larger plants and garden accessories like birdbaths, garden seats, statues, window boxes, garden tools, and accessories. In addition to this, they may sell fertilizer bags, potting mixes, pots, sprayers etc.

1. Tissue culture Units dealing with ornamental plant production-There is a a huge demand for quality supply of ornamental plants, both for commercial cut flowers and pot plants cultivation as well as for landscape industry. Presently, industry supplying tissue culture plants are meagre and a lot of import of floricultural produce is done.
2. Flower Arranging and Décor -The cultural significance of flowers in India, coupled with a growing trend of lavish weddings and events, opens up avenues for entrepreneurs in the domain of flower decor and arrangement services. It's an art that requires creativity, a keen aesthetic sense, and knowledge of different types of flowers. Floral designers can work in or own flower shops, or provide their services for events like weddings, corporate functions, or funerals.The magnificent floral decorations cost range from Rs 10,000 to 20 lac in wedding and religious ceremonies.
3. Wholesale and Retail Floristry- The growth of urban areas and the increase in the standard of living has boosted the demand for ornamental flowers. Entrepreneurs can exploit this demand bysetting up wholesale or retail floristry businesses that cater to individual consumers, event planners, hotels, andcorporate.
4. Flower Exports- There's a high demand for Indian flowers in the global market, especially in countries like the Netherlands, Germany, the UK, and UAE. Entrepreneurs can tap into this potential by establishing export-oriented floriculture units. Indian flowers, especially traditional ones like roses, jasmine, and marigold, are in high demand in international markets. Professionals in this field work on identifying the right markets, ensuring quality control, handling logistics, and understanding international trade laws.
5. Dry flower business-It is budding industry that has good demand in international market as it caters to more than 50 percent share in floriculture exports. The demand for dry flowers is increasing at an impressive rate of 8-10% and therefore there is a great scope for the Indian entrepreneurs. Dry flower products like dry flower bouquets, cards, wall hangings, sceneries, frames, wreathes, pomander etc are popularising in the market. Be-

sides, a significant component of dry flower export comprise of potpourris.Tamil Nadu, Andhra Pradesh and West Bengal have emerged as major hubs of dry flowers industries in recent times.

6. Value addition based industries like essential oils, pigments etc- These are upcoming industries that are still in their nascent stage but has high demand in the inter-national market. Emerging demand in spas and aromatherapy along with growing awareness towards eco-friendliness and natural products has given impetus to essential oils and perfumery from natural sources. According to a report by the National Research Development Corporation (NRDC), the total demand for essential oil is roughly split between the fragrance industry (60%), flavour industry (20%), and pharmaceutical industry (20%). Concern for the environment and health created an interest in natural dyes. Flowers are the main source of colours and there are more than 450 species which yield colours. India is endowed with rich plant wealth and has a great potential to cope with a major portion of the world demand. In food industries floral pigments are mainly used to get natural appealing of the product as well as preservative and even as nutritional supplement (carotenoids) to improve the quality of the produce and also to add the aroma and flavor of food items. They are normally used in Salad dressing, Ice cream, dairy products, other food stuffs with high fat content, Soft drinks, Bakery products, Jams and Confectionary etc. Marigold Xanthophylls / Lutein are mainly used for imparting color in egg yolk, broiler skin and also in fish. Carotenoid composition of marigold flower extract is important as nutritional supplements and increases its value as a poultry feed colourant because it contains more biologically useful lutein compounds. Finely ground blossom meal or the extract, usually saponified for better absorption, is added to the poultry feed. These products are traded as 'Aztec marigold' or 'Marigold meal' and 'Marigold extract' or 'Adoptinol'.
7. Landscape designing and Gardening-As urban spaces crave greenery and beauty, there is a growing demand for landscape designers. Landscape Gardening is the application of garden forms, different styles, methods and materials with a view to improving the landscape. Landscape gardening is presently gaining great importance and developing with the growing awareness for environmental moderation along the developing concept of bioaesthetic planning. The prime application of landscape designing lies in aesthetic developments and modernization of cities, town, villages, roadways, airports, railway stations, railway lines, bus terminus, city parks, educational institutions as well as residential societies, against industrial fast growing pollution. Well planned landscape around building or residence not only adjoins aesthetic value to it but also affix about 30% high value of real estate.
8. Indoor gardening in urban areas- Rapidly growing urbanization and population explosion with hasty life style have created space paucity for living, wherein pot plants offers some relief with a great scope for indoor

gardening. Concept of container gardening, window-box gardening, gardening in hanging containers, dis gardening, miniature gardening, terrariums, etc. has been increasing for office and home.

9. Technology-Driven Flower Businesses-Tech-savvy entrepreneurs can leverage technology to establish online flower delivery platforms, floriculture consultancy services, or agritech solutions for floriculture farming
10. Sub-Industries associated withfloriculture– Subindustries viz., green house construction and installation, shade net manufacturing, grading-packaging machines, pot and planters manufacturing and dealing, growing media processing etc. are also in demand with the rising floriculture industry.
11. Novel areas-edible flowers, flower/aroma therapy- In recent years the interest in the consumption of fresh flowers is renewed and revived, not just because of new demands but with the growing concern of health, immunity and nutrition. Edible flowers are becoming more popular with their use as salads, tea, soups, desserts and drinks. Pansy, nasturtium, marigold and orchids are used in desserts specially cakes and ice creams. Rose has been traditionally used as rose syrup for preparation of summer drink, sugar based gulkand, dried rose as pankhuri- as an ingredient in traditional recipes and as addition in different desserts.
12. Research and Development-The development of new flower varieties, pest management, integrated nutrition management, growing media, post-harvest technology are critical areas where trained researchers are needed. These professionals work in public sector mainly in the agricultural universities of state, state departments, research institutions of ICAR or in private sector with large floriculture companies.

Benefits of Choosing Career in Floriculture

While choosing career path, diverse factors right from skill development to growth perspective are considered. There are some specific added benefits of choosing career in floriculture as mentioned below:

- Opportunities for entrepreneurship and self-employment
- Satisfaction derived from working with nature and nurturing plant life
- Positive impact on the environment and sustainability
- Growing demand for skilled professionals in the industry
- Contribution in creation of aesthetic space
- Nurturing creativity

Factors that Boost the Entrepreneurial Opportunities in Flower Sector in India

India has the potential to develop and flourish the lucrative floriculture industry with several advantages. India is at the edge of advantage, with following merits for strengthening of the flora-industry. Some of the advantages are given below:

- Agro-climatic conditions suitable for various flower crops.
- Proximity to the consuming markets like Europe, Middle East countries and Far East.
- Winters are not very cold but suitable for quality cut flower production without the extra expense of green house heating.
- Free from adverse climatic disturbance like monsoon, snowfalls and rainstorms during the peak demand worldwide from November-March (Christmas, New year day, Valentine's Day etc.).
- Availability of skilled manpower
- Availability of growing and expanding domestic market
- Growing awareness for quality production in nurseries and development of several intensive tissue culture units.
- Constant research on hi-tech water and nutrition management in cut flowers in different research centers, SAUs as well as under Directorate of Floriculture Research.
- Availability and adoptability of sophisticated irrigation systems and climate control systems.
- Availability and adaptability of new modified green house technology suitable for Indian conditions.
- The growing role in Urban greening and urban aesthetics of floriculture
- Application of flowers in pharmaceuticals and aroma therapy
- Growing demand for Pot plants in urban areas
- Landscape architecture as a business potential specially for environmental mitigation
- Availability of expert consultancy in protected cultivation suitable to our country.
- Availability of skilled manpower through skill development programs
- Govt. policies are export friendly and floriculture is one of the thrust area for promotion of exports.
- High employment generation and entrepreneurship development opportunities
- Span of urban greening for rooftop gardens, balcony gardens, vertical greening

- Need of theme based landscaping for industries, residential areas and corporate sector
- Indoorscaping for improving indoor air quality in urban vicinity
- High potential for environmental services through carbon sequestration and pollution control in peri-urban areas
- Huge opportunities in landscape business for town planning, turf management, industries, sports complex etc.

Government Support in Floriculture Sector

Government of India has identified floriculture as a sunrise industry, and accorded it 100% export oriented status. Owing to the steady increase in demand for flowers, floriculture has become one of the important Commercial trades in Agriculture. Hence commercial floriculture has emerged as a hi-tech activity-taking place under controlled climatic conditions inside a greenhouse. Floriculture in India is being viewed as a high growth Industry. Commercial floriculture is becoming important from the export angle.

Many developmental programmes through different schemes of National Horticulture Board (NHB), National Horticulture Mission (NHM), Mission for Integrated Development in Horticulture (MIDH), Horticulture Mission for North East & Himalayan States etc., under the aegis of Ministry of Agriculture have been playing a central role in uplifting floriculture growth in the country. National Bank for Agricultural Rural Development (NABARD) and Rashtriya Krishi Vikas Yojana (RKVY) have also been creating awareness about various components of horticulture including floriculture in rural areas as well as providing financial assistance to the farmer to adopt different technical avenues for flower cultivation. Furthermore, the Government of India (GoI) is undertaking numerous initiatives to promote export-oriented businesses by providing income-tax holidays to new floriculture entrepreneurs across the country. This, along with the subsidiary policies offered by the Agricultural and Processed Food Products Export Development Authority (APEDA) and Mission for Integrated Development in Horticulture (MIDH)for the adoption of and improvements in cold storage facilities, refrigerated vans and pre-cooling units along with the greenhouse units is expected to provide an impetus to the market growth further.

Research activities on floriculture are being carried out at several research institutions under the Indian Council of Agricultural Research (ICAR), State agriculture universities and under the Directorate of Floricultural Research, Pune. Various training related programmes for farmers and entrepreneurs are also being conducted for tot. Recently, training programs for training of gardener, florist, protected cultivation expert, etc under Agriculture Skill Council of India (ASCI) have been initiated to create technically skilled manpower in the particular field. Further, Krishi- rath and Krishi mahotsava program have added to the development of flower sector throughout the country. Government offers a range of concession on seeds, planting materials, various types of equipment and airfreight for export. It is anticipated that improved policies would enhance floriculture in world trade.

Further, some remarkable developments in Floriculture like development of Model Floriculture Centre and Horticulture Training Centre in Maharashtra, Floriculture Infrastructure Park in Tamil Nadu, strengthened Flower grower association and identification and establishment of Agri export zone for floriculture by APEDA in Sikkim, Tamil Nadu, Uttranchal, Karnataka and Maharashtra have further given boost to the Flora Industry.

Skills to Succeed in Floriculture Career

Students need to develop the following skills to excel in floriculture sector

- A strong foundation in plant biology, horticulture, and related sciences. Knowledge about basics in soil science, Plant identification, plant categorization as per the weather, media, growth habit, flowering habit, etc is necessary.
- Practical experience in plant propagation, plant cultivation, crop management and maintenance. These skills will boost the confidence in the students as this business deals with live plants.
- Creativity and artistic ability for design and arrangement. This important to keep the floriculture based business flourishing.
- Effective communication and interpersonal skills for customer service and teamwork.
- Business acumen and entrepreneurial mindset for management roles.

Education and Skill Requirements

While a basic love for nature and an eye for beauty are intrinsic requirements for a career in floriculture, specific roles might require specialized training or education. For instance, a degree of Bachelors in horticulture would be helpful for those interested in flower farming or flower business. Further, one can go ahead with in depth study with Masters and PhD in Floriculture and landscaping. For florists, training or certificate course in flower arrangement and design would be beneficial. As for exports or landscape design, nursery business, appropriate vocational or professional training is advisable.

Companies Excelling in Floriculture

The **Indian floriculture market** is expected to reach a value of INR 460.6 Billion by 2028, exhibiting a growth rate (CAGR) of 13.1% during 2023-2028, as per the report of IMARC Group. Private sector is leading the floriculture industry with excellence. The leading **floriculture companies in India** are increasingly focusing on shifting toward virtualized trade networks to provide better logistics coordination with improved information availability in the floriculture supply chain. This allocates all the stakeholders involved in the supply chain to access real-time available information, which further aids in ensuring smoother, faster and cost-effective deliveries of perishable and degradable floriculture products. Top leading floriculture companies are listed below:

1. Karuturi Global Limited

Karuturi Global Limited is the leading producer and exporter of cut roses on the global level. Incorporated in 1994, with headquarter at Bengaluru, it has operations spread across India, Ethiopia, and Kenya. It produces around 500 million rose stems annually with a capacity to process 12 million premium cut roses at its state-of-the-art facilities. It has also set up a wholly-owned subsidiary named Ethiopian Meadows Plc in Ethiopia, Africa that focuses on the production of HT roses. Along with this, Karuturi Global Limited also supplies other cut rose products, such as rose plants, coco peat and coco cups, to their customers across numerous countries, including Germany, the United Kingdom, Italy, Singapore, Taiwan, Australia, Japan, New Zealand and across North American region.

2. Syngenta Flowers

Syngenta is one of the largest science-based agtech companies worldwide, established in 2009. Headquartered in Switzerland, it employs more than 26,000 individuals across 100 countries. It is increasingly focusing on sustainable practices for enhancing biodiversity, improving soil and creating an impact on the global level. The company is also a global leader in sustainable flower production. With a large team of experts on advanced genetics and breeding, Syngenta offers retailers, distributors and professional growers across 75 different countries with state-of-the-art seeds, young plants and cuttings, along with training and expert advice.

3. Ferns and Petals Group

Ferns and Petals Group, established in 1994 with headquarter in New Delhi, is one of the largest flower and gifts retailers in India. The company is regarded as the preferred gifting solution provider amongst the masses that has provided fresh flowers and thoughtful gifts to their customers across the country for over 25 years. It has now expanded as the largest florist chain in India with over 320 stores across all major cities and towns of the country. It has also successfully expanded its international footprints into other parts of the world, particularly the South East Asian and Middle East markets. It has 3 stores in the UAE and 1 store in Singapore, along with a strong online presence with separate websites for each of the territories.

4. Florance Flora

Florance Flora, established in 1980 at Bengaluru, is a privately-owned supplier of young plants for commercial cut flowers and pot plants production with an enhanced focus on customer satisfaction.. One of the leading floriculture companies in India, it has a vast crop portfolio and collaborations with suppliers around the world. It provides numerous services to its customers, including plant inspection, crop selection, etc.

5. Soex Flora

Soex Flora, established in 2009 at Mumbai, is one of the leading growers and exporters of fresh cut flowers across India. The company offers a vast range of rose varieties in its greenhouses that are powered by innovative and latest technologies. It comprises a highly qualified team of experienced floriculturists and plant technologists from across the country that continually ensures high-quality yield of roses throughout the year. It also has the capacity to deliver its services across the globe in a limited period of time.

Other big players in the floriculture industry include The Kariki Group, Dummen Orange, Oserian Development Co Ltd., Selecta Klemm GmbH & Co., Marginpar BV, Ruprelia Group, Multifloa Ltd., Karen Roses etc.

The flourishing floriculture in India has been attracting startups also. Rose Bazaar is one such start up that is based at Bengaluru, that is offering subscription based model for flower supply. With a monthly delivery over 15000 volume, it is well positioned to become a major player in the floriculture industry in India. A few companies have been excelling in dry flower business also. Ramesh Flowers Pvt. Ltd., Fauna International, W.Hogewoning and Sibaflor Natural Decorations at Thoothukudi (formerly Tuticorin), Tamil Nadu and Salem Spices Pvt Ltd., Salem, Tamil Nadu are the major players in dry flower business in India and exporting dry flowers all over the world from Europe to US and from South Africa to Australia.

Challenges in Floriculture Career Path

Despite its potential, a career in floriculture also presents some challenges which needs to be encountered well and overcome.

- Perishable commodity
- Vast effect of climate and weather condition
- Seasonal fluctuations in demand
- Wide range of commodity and specific demand for particular range
- Artificial flowers have been capturing the market
- Need for continuous learning and adaptation to new technologies and trends
- Competition from large-scale commercial

Conclusion

Flora industry has great export potential which can improve national economy as well as beautify our country with vibrant colours and clean green environment. Floriculture offers students a fulfilling and dynamic career path, with ample opportunities for growth, creativity and environmental stewardship. By acquiring the necessary skills and education, and embracing innovation and adaptability, students can thrive in this promising field and contribute to a more sustainable, beautiful, and vibrant world.A career in floriculture can be rewarding - not just financially, but also emotionally. After all, it's a field that allows one to work with nature's beauty and bring smiles to people's faces. With the right education, skills, and a love for flowers, one can build a flourishing career in this vibrant sector.

The verdant vistas of opportunity in the floriculture sector, fueled by India's rich floral diversity and growing demand, hold immense entrepreneurial potential. With government support and an increasing global demand for Indian flowers, the prospects for floriculture entrepreneurship in India are vibrant and full of promise.

Index

9 789359 192437

Printed at Repro India Ltd.